D1196907

Categories, Bundles and Spacetime Topology

Mathematics and Its Applications

Managing Editor:

M. HAZEWINKEL

Centre for Mathematics and Computer Science, Amsterdam, The Netherlands

Editorial Board:

F. CALOGERO, *Università degli Studi di Roma, Italy*
Yu. I. MANIN, *Steklov Institute of Mathematics, Moscow, U.S.S.R.*
A. H. G. RINNOOY KAN, *Erasmus University, Rotterdam, The Netherlands*
G.-C. ROTA, *M.I.T., Cambridge, Mass., U.S.A.*

C. T. J. Dodson

Department of Mathematics, University of Lancaster,
Lancaster, United Kingdom

Categories, Bundles and Spacetime Topology

Second, Revised Enlarged Edition

KLUWER ACADEMIC PUBLISHERS

DORDRECHT / BOSTON / LONDON

Library of Congress Cataloging in Publication Data

Dodson, C. T. J.
   Categories, bundles, and spacetime topology / C.T.J. Dodson. --
2nd, rev., enl. ed.
      p.   cm. -- (Mathematics and its applications)
   Bibliography: p.
   Includes index.
   ISBN 9027727716
   1. Topology. 2. Fiber bundles (Mathematics) 3. Vector bundles.
   4. Categories (Mathematics)   I. Title. II. Series: Mathematics and
   its applications (D. Reidel Publishing Company)
   QA611.D58 1988
   514--dc19   88-15566
                                                              CIP

ISBN 90-277-2771-6

Published by Kluwer Academic Publishers,
P.O. Box 17, 3300 AA Dordrecht, The Netherlands.

Kluwer Academic Publishers incorporates
the publishing programmes of
D. Reidel, Martinus Nijhoff, Dr W. Junk and MTP Press.

Sold and distributed in the U.S.A. and Canada
by Kluwer Academic Publishers,
101 Philip Drive, Norwell, MA 02061, U.S.A.

In all other countries, sold and distributed
by Kluwer Academic Publishers Group,
P.O. Box 322, 3300 AH Dordrecht, The Netherlands.

First edition of this book was published by
Shiva Publishing Limited.

Printed in The Netherlands

To

Ben Radvan

&

Heinz   Corte

*For introducing me to research
and for their guidance over the
ten years 1959-1969 , when we
worked together on statistical
geometry.*

SERIES EDITOR'S PREFACE

Approach your problems from the right end
and begin with the answers. Then one day,
perhaps you will find the final question.

'The Hermit Clad in Crane Feathers' in R.
van Gulik's *The Chinese Maze Murders*.

It isn't that they can't see the solution. It is
that they can't see the problem.

G.K. Chesterton. *The Scandal of Father
Brown* 'The point of a Pin'.

Growing specialization and diversification have brought a host of monographs and textbooks on increasingly specialized topics. However, the "tree" of knowledge of mathematics and related fields does not grow only by putting forth new branches. It also happens, quite often in fact, that branches which were thought to be completely disparate are suddenly seen to be related.

Further, the kind and level of sophistication of mathematics applied in various sciences has changed drastically in recent years: measure theory is used (non-trivially) in regional and theoretical economics; algebraic geometry interacts with physics; the Minkowsky lemma, coding theory and the structure of water meet one another in packing and covering theory; quantum fields, crystal defects and mathematical programming profit from homotopy theory; Lie algebras are relevant to filtering; and prediction and electrical engineering can use Stein spaces. And in addition to this there are such new emerging subdisciplines as "experimental mathematics", "CFD", "completely integrable systems", "chaos, synergetics and large-scale order", which are almost impossible to fit into the existing classification schemes. They draw upon widely different sections of mathematics. This programme, Mathematics and Its Applications, is devoted to new emerging (sub)disciplines and to such (new) interrelations as exempla gratia:

- a central concept which plays an important role in several different mathematical and/or scientific specialized areas;
- new applications of the results and ideas from one area of scientific endeavour into another;
- influences which the results, problems and concepts of one field of enquiry have and have had on the development of another.

The Mathematics and Its Applications programme tries to make available a careful selection of books which fit the philosophy outlined above. With such books, which are stimulating rather than definitive, intriguing rather than encyclopaedic, we hope to contribute something towards better communication among the practitioners in diversified fields.

There are by now several annual and bi-annual conferences devoted to 'differential geometric methods in physics', mostly complete with regular proceedings. It has simply become clear that differential topology and geometry (and algebraic geometry) provide *the* natural setting for a great deal of modern physics especially general relativity and gauge theories, which does not mean that the use of the languages and techniques is limited to these chunks of physics.

Especially relevant are manifolds with extra structure mostly given in terms of bundles. That bundles (in mathematics) and gauge fields (in physics) arose independently, and initially evolved without contact, is of historical interest. They have now been recognized as the same structure - a gauge field is a section in a bundle - to the delight of mathematicians and physicists both. And thus the age-old partnerships between mathematics and physics have been renewed and strengthened.

Working with manifolds, bundles, and sections instead of Euclidean spaces and functions forces one to think and work much more coordinate freely (i.e. reference frame independently), one learns to distinguish between different kinds of objects (even though, locally, in a given coordinate system, they are all given by the same number of functions), and one rapidly learns that some constructions are more natural and better behaved than others. This is where category theory comes in; a theory, which is a great help in keeping track of what is going on. There is a lot of very sophisticated and powerful category theory. However, the part needed by all mathematicians and physicists, termed naive category theory by the present author, is small, easily absorbed, (and virtually indispensable); it is mostly a language with grammar and contains no hard theorems. Here it takes up no more than some 35 unhurried pages. Also the underlying ideas have, philosophically speaking, much in common with such basic ideas as covariance and symmetry in physics.

'Structures on manifolds' are today, 1988, still more important then they were in 1980 when the first version of this book appeared. 'Strings' contributed greatly to that. There has appeared a growing need for books which treat the topic well and understandably. Thus I am most pleased to have in the series this 'royal road introduction' to the topic, with special emphasis on space-time structures.

The unreasonable effectiveness of mathematics in science ...

    Eugene Wigner

Well, if you know of a better 'ole, go to it.

    Bruce Bairnsfather

What is now proved was once only imagined.

    William Blake

As long as algebra and geometry proceeded along separate paths, their advance was slow and their applications limited.

But when these sciences joined company they drew from each other fresh vitality and thenceforward marched on at a rapid pace towards perfection.

Joseph Louis Lagrange.

Bussum, May 1988   Michiel Hazewinkel

Preface to the Second Edition

The main change from the First Edition is the addition of a new section, §6 in Chapter V. This gives an account of some recent developments pertaining to stability of geometrical singularities in spacetime. The new theory of systems of connections and their universal connections is outlined for this purpose, and may prove to be useful in other contexts for field theories in physics.

A number of errors in the earlier edition have been corrected, and a supplementary bibliography is provided.

I am grateful to colleagues for comments and suggestions, to Dr David Larner of Kluwer for his encouragement and efficiency and to my wife Frances for her assistance in preparing the revised manuscript.

Kit Dodson
Lancaster, January 1988

Contents

Introduction

This book grew out of assorted lectures, short courses and seminars
I have given at Lancaster, Trieste and elsewhere between 1975 and
1979.   The choice of material has been largely influenced by my
interest in spacetime geometry and by the interests of many, valued
friends working in various branches of theoretical physics.

One of the principal mathematical tools of physicists is
differential geometry and conversely much benefit has been derived
in mathematics from considerations of the formulation of physical
theories.   Perhaps the most important current topic is bundle
theory, because it is now enjoying a central role in both classical
relativity and gauge theories.   This is reflected in the amount of
space devoted to structures on manifolds in Chapter IV, about one
third of the book.   To save the time of those occupied with
applications, Chapter IV begins with a quick reminder of the
elementary aspects of manifolds.

Once hooked on bundle theory, a theoretical physicist soon
finds that the intricate mixture of algebra and analysis on
manifolds spawns vast families of maps going hither and thither
among his spaces.   Many of these maps turn out to have excellent
pedigrees : they are canonical, or natural entities and they give
rise to standard, or universal constructions.   The right way to
view it all is through the theory of categories, and increasingly
the language of this theory is appearing in the literature of
physics.   Chapter II is an introductory course on category theory
with many examples, but little in the way of proofs since these are
available in the more advanced texts to which, as elsewhere in the
book, abundant references are given.   The most widely used

category in theoretical physics is Top, consisting of topological
spaces and continuous maps.   For, most experiments are unable to
detect discontinuities and so we resort to simple models from
topology : small changes have small effects.   The category Top
is studied in Chapter III.   There, after a short recounting of
basic ideas, fully detailed proofs are given of the existence and
uniqueness of limiting topologies on a given set under various
conditions that commonly arise in applications.   Thus, the
completeness of Top as a category is displayed by construction of
products, coproducts, equalizers and coequalizers.

     The last part of the book, Chapter V, is a survey of certain
global features of spacetime 4-manifolds.   There is little here
that is actually new except in emphasis, but the material was
fragmented rather widely through journals and conference proceedings
so it seemed appropriate to bring it together in one place.   We
consider first the existence of Lorentz structures and how they
relate to paracompactness and nowhere-zero vector fields.   Next
comes orientability and its relation to nowhere-zero differential
4-forms.   Orientability is effectively always assured because we
can use a covering spacetime, and we see also the relation to time-
and space-orientability.   The correspondence between spinor
structures and parallelizability is outlined and we finish with an
account of the geometrical definition and handling of singularities,
by bundle completion and modifications thereof.

     The book is organized with a simple numbering system for its
subsections, facilitating the copious cross references; there is
a comprehensive index and a detailed bibliography.   Standard
abbreviations and conventions are described in Chapter I.
Throughout, exercises are disguised as examples and omitted proofs;
the omission of a proof is always accompanied by a reference to
its location and often also by a hint at the construction.

     As is usually the case, I am indebted to more people than I
can name here for help at one stage or another with this book.
However, I have benefited particularly from discussions with
L.W. Flinn, L.J. Sulley, M.J. Slupinski, M. Radivoiovici, C.J.S.
Clarke and R.W. Tucker; also, while at the International Centre

For Theoretical Physics, Trieste, I received much encouragement
from its Director, Abdus Salam; to all these and to the others
with whom I have had discussions at conferences and seminars go
my sincere thanks.   I wish finally to thank Lin Sulley and David
Towers for help in checking the draft, Alexandra Dodson for help
with the bibliography and index, and Sylvia Brennan for yet another
excellent job of typing.

Kit Dodson

Lancaster, January 1980.

I Preliminaries

Though elementary definitions from topology and differential
geometry have been provided for convenience, some familiarity with
them will be expected.   Also, some experience of sets, vector
spaces and basic group theory will be assumed, but the overall
algebraic content of our material is slight and the treatment of
categories is from first principles.   Our main reference text for
prerequisite material is the book Tensor Geometry [28] , which
gives a leisurely development from vector spaces through topology
to curved manifolds and relativistic spacetime.

It is intended that the book be read in the order of its
contents.   However, someone who needs a quick grasp of bundle
theory could go directly to Chapter IV and ignore references to
categorical constructions that are not familiar.   The spacetime
structure in Chapter V relies on material in Chapter IV and the
study of the topological category in Chapter III depends on the
category theory in Chapter II.

Wherever it seems to help, without undue loss of precision,
we shall adopt an intuitive viewpoint in our constructions.   So,
for a start, we shall assume a common language of elementary set
theory.   For a splendidly readable summary of set theory see the
first part of Devlin [21], and for an appreciation of why set
theory is still a lively developing subject read the second part
too.   It will usually be sufficient for our purposes to assume
that our sets belong to some fixed universe of sets; but we also
need proper classes (cf. [46]).   We shall make use of the usual
logical symbols and abbreviations as follows:

$x \in X$	:	"x an element of the set X"
$\forall x$	:	"for all x"
$\exists x$	:	"there exists x"
$\exists! x$	:	"there exists a unique x"
$\not\exists x$	:	"there does not exist x"
$\Longrightarrow$	:	"implies" or "then"
$\Longleftrightarrow$	:	"if and only if" or "is equivalent to"
$\emptyset$	:	"the empty set" or "the set with no elements"
$A \subseteq X$	:	"$(\forall x \in A)(x \in X)$" or "A a subset of X"
$A \not\subseteq X$	:	"$(A \subseteq X)$ is false"
$A = X$	:	"$(A \subseteq X)$ and $(X \subseteq A)$"
$A \subset X$	:	"$(A \subseteq X)$ and $(X \neq A)$"
$\{x \in X \mid P(x)\}$	:	"the set of elements of X for which the statement P is true"
$X \backslash A$	:	"the complement of A in X" $\Longleftrightarrow$ "$\{x \in X \mid x \notin A\}$"
$A \cap B$	:	"$\{x \in X \mid (x \in A) \text{ and } (x \in B)\}$; $A, B \subseteq X$"
$A \cup B$	:	"$\{x \in X \mid (x \in A) \text{ or } (x \in B) \text{ or both}\}$; $A, B \subseteq X$"
$f : X \to Y$	:	"$(\forall x \in X)(\exists! \; f(x) \in Y)$" $\Longleftrightarrow$ "f is a map from set X
$: x \mapsto f(x)$		to set Y, sending typical x to f(x)"
dom f	:	"the <u>domain</u> of f, X when $f : X \to Y$"
cod f	:	"the <u>codomain</u> of f, Y when $f : X \to Y$"
im f	:	"the <u>image</u> of f, $\{f(x) \mid x \in \text{dom } f\}$"
sub X	:	"the <u>power set</u> of X, $\{A \subseteq X\}$"
$f^{\leftarrow}$	:	"the <u>inverse image map</u> for subsets of cod f; when $f : X \to Y$ then $f^{\leftarrow} : \text{sub } Y \to \text{sub } X : B \mapsto \{x \in X \mid f(x) \in B\}$"
$i : A \hookrightarrow X$	:	"the <u>inclusion</u> map of a subset $A \subset X$"
I_A	:	"the <u>identity</u> map on A"
$f\vert_A$	:	"the <u>restriction</u> of a map f to a subset A of dom f"

X/~   :   "the quotient of a set X by an equivalence

                relation ~ "

[x]$_\sim$:   "the equivalence class containing x "

 $\simeq$:   "is homeomorphic to"

 $\equiv$:   "is equivalent or isomorphic to"

N,Z,R,C :   "natural, integer, real complex number systems"

 $\square$:   "end (or omission) of proof"

A handy reference for sets, functions, relations and their
applications is Birkhoff and Bartree [5].

II Naive category theory

Our objective in this chapter is to assemble some basics of the
theory and to acquire some practice with the language of categories.
There are several very good introductory texts for mathematicians
and the ones by MacLane [67] and Herrlich and Strecker [46] are
probably the best references for physicists who want more details
than we can provide here.   Note that we use the term <u>set</u> where
MacLane uses the term <u>small set</u>: to mean a <u>class</u> that is not a
<u>proper class</u> (cf. [46]).   The important technical point is that
we must not speak of the 'set of all sets' or 'the set of all
groups' these are not constructible from the operations of standard
set theory; they are examples of <u>proper</u> classes.   We shall
suppose that any group, ring, field, vector space or topological
space is a <u>set</u> together with some extra structure, that is a class
which is not a proper class.   Occasionally we shall need to use
maps between classes, then we use the same notation as for maps
between sets.

1.   CATEGORIES AS STRUCTURED GRAPHS

After the natural number system it could be argued that the most
widely applied concept from mathematics is that of a graph.   In
physics we are concerned with the representation of <u>observable</u>
phenomena and this is in principle possible by means of finite
graphs.   The vertices are observable entities of interest and the
edges can represent interactions or motions that may be qualitative
or, at best, quantitative with finite precision.   When we come to
fitting observations in a behavioural model the method is to cloak
their graphical skeleton, usually with <u>continuum</u> theories because

4

at most levels of experimentation real phenomena seem to suffer
small changes when the ambient conditions undergo small changes.
(The classical exceptions of course are changes of state and
fracture phenomena where a continuous input of energy can yield a
discontinuous change in behaviour.   Such exceptional behaviour,
and not only from physics, is currently an active area of application
for catastrophe theory, cf. Poston and Stewart [79]).   So for
much of theoretical physics, <u>topology</u>, the study of continuous
processes, is pre-eminent.   However, the graphs come back again
in the form of superstructure, such as group symmetries, on the
topological spaces.   In mathematics, the concept of graph enjoys
ubiquity.   The facet which we use in the sequel is category theory,
which allows us to abstract similarities in the application of
different mathematical constructions and to exploit a powerful
common language.   The increasing sophistication of studies in
spacetime geometry and associated field theories using, for example,
bundle theory and algebraic topology/geometry gradually is feeding
into physics some of the categorical language.   Inevitably, more
category theory will be used at the formulation stage of physical
theories and, there, care will be needed to avoid unnecessary
obscurity so in this section we collect some basic terms, with
examples, intended to give an accurate though not detailed
impression of the ingredients of category theory.   Subsequently
we shall consider some commonly occuring construction processes
that are essentially categorical.

1.1   *Graph definitions*

A <u>graph</u> is a collection ⊡ of <u>objects</u> and a collection ⊞ of
<u>arrows</u> with two operations, associating with each arrow a unique
pair of objects, dom(a) and cod(a);   that is to say,

$$\text{dom} : ⊞ \to ⊡, \quad \text{cod} : ⊞ \to ⊡ ,$$

are <u>maps</u>.   Actually this is a <u>directed</u> graph (cf. Higgins [47])
but we shall need no other kind.   We wish to allow collections to
be sets <u>or</u> proper classes (cf. [46]).

Example

The prototype of all graphs is that for transport systems, such as the London Underground. Then ⊡ is the set of stations and ⊞ is the set of train routes between stations. Two elements of ⊡ are Whitechapel and Stepney Green and there are just two arrows from the first to the second. These arrows are m_o (for Metropolitan Line) and d_o (for District Line) so we represent this situation locally by the diagram

$$\text{Whitechapel} \quad \overset{m_o}{\underset{d_o}{\rightrightarrows}} \quad \text{Stepney Green}$$

Here we have
$$\text{dom}(m_o) \;=\; \text{dom}(d_o) \;=\; \text{Whitechapel}$$
$$\text{cod}(m_o) \;=\; \text{cod}(d_o) \;=\; \text{Stepney Green}$$

There are of course many more ways by Underground from Whitechapel to Stepney Green. The next shortest after the above pair is a composition of five arrows $m_3 \circ c_2 \circ c_1 \circ m_2 \circ m_1$:

$$\begin{array}{ccccc}
\text{Liverpool} & \xrightarrow{c_1} & \text{Bethnal Green} & \xrightarrow{c_2} & \text{Mile End} \\
\text{Street} & & & & \\
\uparrow m_2 & & & & \downarrow m_3 \\
\text{Aldgate} & & & & \\
\text{East} & \xleftarrow{m_1} & \text{Whitechapel} & \text{-- } \to \text{ --} & \text{Stepney Green}
\end{array}$$

London Transport calls this composite way an available route and so we could include it in the set ⊞ . Plainly, we can only compose contiguous arrows and since the idea of Underground maps is to help the traveller, it is clearer to leave the traveller to do his own compositions. Hence, route maps only show the normally nonstopping routes between stations.

1.2   *Category axioms*

A <u>category</u> is a graph in which :

(i)   every object A has an <u>identity</u> arrow I_A with
        dom I_A = cod I_A = A;

(ii)   there is the possibility of <u>composition</u> of arrows in the
        sense that given any arrows A $\xrightarrow{f}$ B and C $\xrightarrow{g}$ D then
        they <u>compose</u> to give a unique arrow A $\xrightarrow{g \circ f}$ D if and
        only if B = C; if B ≠ C then g∘f is not defined;

(iii) wherever it is defined, composition is <u>associative</u>:
        (g∘f)∘h = g∘(f∘h), (cf. §1.6 Ex.1);

(iv)   identity arrows always compose with A $\xrightarrow{f}$ B to give
        $I_B \circ f = f$ and $f \circ I_A = f$, (cf.§1.6 Ex.2).

In a category, the arrows are usually called <u>morphisms</u>.

Remark

As a refinement of the concept of a graph these defining properties
are fairly natural; associativity of set-theoretic maps prompts
(iii).   An elementary algebra with a non-associative composition
is that of the Cayley numbers or octonions (cf. Porteous [78]) and
we shall make use in the sequel of Lie algebras, which are also
non-associative.

Example 1

The categories most commonly met in applications have objects that
are structured <u>sets</u> and morphisms .that are structure-preserving
<u>maps</u> with the usual map composition.   Among these are:

Grp,   consisting of groups and group homomorphisms;
Ab ,   consisting of Abelian groups and group homomorphisms;
Vec,   consisting of vector spaces and linear maps;
Top,   consisting of topological spaces and continuous maps.

Example 2

A group (G,∘) is a category with one object (the underlying set),
and each element is an arrow (actually a map from the set of
elements to itself); also each arrow is <u>invertible</u>:

$$(\forall x \in G)\,(\exists x^{-1} \in G) \;:\; x \circ x^{-1} = x^{-1} \circ x = e = I_G.$$

Example 3

A category in which every arrow is invertible, that is :

$$(\forall\; A \xrightarrow{\;f\;} B)\,(\exists\; B \xrightarrow{\;f^{-1}\;} A) \;:\; f \circ f^{-1} = I_B, \; f^{-1} \circ f = I_A,$$

is called a __groupoid__.   These are studied in detail by Higgins [47].

Example 4

Set is the category whose objects are all sets and whose arrows
are all __maps__ among them.   This is an example of a __large__ category:
its collection of objects and arrows is not a set, it is a proper
class.   A category is called __small__ if it is not large.

Example 5

We can make a larger category than Set by using the same objects
but allowing all __binary relations__ to be morphisms.   This category
is denoted Rel.

Example 6

Given a category C the __dual__ or __opposite__ __category__ C^{op} is
defined to have the same objects as C but for each morphism

$$A \xrightarrow{\;f\;} B \quad \text{in} \quad C$$

we get an opposite morphism, by interchanging the dom and cod
maps,

$$A \xleftarrow{\;f^{op}\;} B \quad \text{in} \quad C^{op}$$

and composition in C^{op} is defined by

$$f^{op} \circ g^{op} = (g \circ f)^{op}. \quad \text{(cf. §1.4 Ex.1)}$$

Example 7

Given categories C_1, C_2 their __product category__ $C_1 \times C_2$ has as
objects all ordered pairs of objects from C_1 and C_2 and
morphisms are ordered pairs of morphisms from C_1 and C_2.

1.3   *Covariant functor*

A covariant functor F from a category C_1 to a category C_2 is
a pair of maps

$$
F: \begin{cases} F_{obj} & : \quad \text{Objects of } C_1 \longrightarrow \text{ Objects of } C_2 \\[2ex] F_{mor} & : \quad \text{Morphisms of } C_1 \longrightarrow \text{ Morphisms of } C_2 \end{cases}
$$

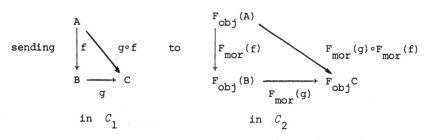

sending ... to ...

in C_1 in C_2

and respecting composition and identities :

$$F_{mor}(g \circ f) \quad = \quad F_{mor}(g) \circ F_{mor}(f)$$

$$F_{mor}(I_A) \quad = \quad I_{F_{obj}(A)}$$

Usually we just refer to covariant functors as <u>functors</u>, and also
we omit the subscripts from F_{obj} and F_{mor} since it should be
clear whether we are acting on an object or a morphism.   The
functor F is called : <u>faithful</u> if F_{mor} is injective; <u>full</u>
when F_{mor} is surjective onto the C_2-morphisms among objects in
the image of F_{obj}; an <u>isomorphism</u> if both F_{obj} and F_{mor} are
bijective.

Example 1

The power set functor from Set to Set is defined by

$$
P : f \quad \begin{array}{c} A \\ \downarrow \\ B \end{array} \quad \longmapsto \quad \begin{array}{c} \text{Sub A} \\ \downarrow f_{\rightarrow} \\ \text{Sub B} \end{array}
$$

where,

$$f_{\rightarrow} : \text{Sub A} \longrightarrow \text{Sub B} : S \longmapsto \{f(x) \in B \mid x \in S\}.$$

It is customary to omit the subscript from $f_{\rightarrow}$.

Example 2

A category C is called <u>concrete</u> if there is a faithful functor F
from C to Set, then F is called a <u>forgetful functor</u>, since it
simply forgets the extra structure in C.   Similarly, forgetful
functors arise between other categories.

Example 3

Composites of full/faithful functors are also full/faithful functors.

Example 4

The categories Grp, Ab, Vec and Top are concrete categories.

Example 5

The set of all (small) categories with functors among them forms a
large category, Cat.

1.4   *Contravariant functor*

A <u>contravariant functor</u> F from a category C_1 to a category C_2
is a pair of maps

$$
F : \begin{cases} F_{obj} & : \text{ Objects of } C_1 \longrightarrow \text{ Objects of } C_2 \\[2ex] F_{mor} & : \text{ Morphisms of } C_1 \longrightarrow \text{ Morphisms of } C_2 \end{cases}
$$

sending
$$
\begin{array}{ccc}
A & & F_{obj}(A) \\
\downarrow f & \text{to} & \uparrow F_{mor}(f) \\
B & & F_{obj}(B)
\end{array}
$$

in C_1 in C_2

and respecting identities but reversing compositions :

$$
F_{mor}(g \circ f) = F_{mor}(f) \circ F_{mor}(g)
$$
$$
F_{mor}(I_A) = I_{F_{obj}(A)} .
$$

We do not drop the qualification 'contravariant' when referring to
these entities, though it is usual to drop the subscripts to the

maps F_{mor}, F_{obj}.

Remark

The origin of the term contravariant is plain: we see that, in the
image, morphisms are reversed with respect to objects.

Example 1

The dualising operation (§1.2 Ex.6) is a contravariant functor
from any category to its opposite.

Example 2

Real vector spaces and linear maps among them form a large category.
R-Vec.   Every vector space V in this category has a dual space
$V* = L(V;R)$, consisting of real valued linear maps from V to R.
Also, for every morphism $f : V \longrightarrow W$ in R-Vec there is a dual
linear map

$$f* : \quad W* \longrightarrow V* : \quad x \longmapsto x \circ f.$$

Hence we obtain a contravariant functor from R-Vec to itself by
taking duals :

$$
\begin{array}{ccc}
V & & V* \\
f \downarrow & \longmapsto & \uparrow f* \\
W & & W*
\end{array}
$$

Evidently R can be replaced by any field and we could equally
well work with modules over any ring.

Example 3

There is a <u>contravariant power set functor</u> from Set to Set given
by

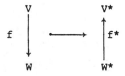

$$
P* : \quad
\begin{array}{ccc}
A & & \text{sub } A \\
f \downarrow & \longmapsto & \uparrow \overleftarrow{f} \\
B & & \text{sub } B
\end{array}
$$

where, as before, $\overleftarrow{f}$ is the inverse image map on subsets:

$$\overleftarrow{f} : \quad \text{sub } B \longrightarrow \text{sub } A : S \longmapsto \{x \in A \mid f(x) \in S\}.$$

As with the term contravariant, we cannot afford to omit the
superscript from $f^{\leftarrow}$, it reminds us of the backwards going effect
of contravariance on morphisms.

1.5   *Subcategory*

We call a category C_o a <u>subcategory</u> of a category C if the
objects and morphisms of C_o form a subclass of those of C and
the inclusion maps

$$F_{obj} : \text{Objects of } C_o \hookrightarrow \text{ Objects of } C : A \longmapsto A$$

$$F_{mor} : \text{Morphisms of } C_o \hookrightarrow \text{ Morphisms of } C : f \longmapsto f$$

form a functor, the <u>inclusion functor</u>.   We say that C_o is a <u>full</u>
<u>subcategory</u> of C when the inclusion functor exists and is full.

Example 1

The category of all finite sets and maps among them is a full
subcategory of Set; the category of all Hausdorff spaces is a
full subcategory of the (large) category Top of topological spaces
with continuous maps among them.

Example 2

The category Ab of all Abelian groups and homomorphisms among
them is a (full) subcategory of Grp the (large) category of all
groups and homomorphisms among them.

Example 3

Topological metric spaces and isometries (distance preserving maps)
among them form a subcategory of Top.   However, the inclusion is
not full because even though every topological space can be given
at least the trivial metric

$$d(x,y) = 1 \text{ if } x \neq y, \quad d(x,x) = 0 ,$$

not every continuous map need be an isometry in any available
metric.

1.6   *Diagram*

A diagram is rather tricky to <u>define</u>, because of the freedom we
normally wish to exercise in viewing as equivalent, different
representations of a given collection of objects and morphisms.
For convenience we use :

A <u>diagram</u> in a category C is a subclass Δ of C with
specified objects $\{A_i | i \in J_o\}$ and morphisms

$$\{a_{ij}^{k} : A_i \longrightarrow A_j | k \in J_m, \ i, \ j \in J_o\}$$

for some indexing sets J_o and J_m, $J_o \neq \emptyset$. To avoid technical
difficulties, we shall agree to interpret this definition with a
spirit of reasonableness.

Remark

Most of the applications of category theory are concerned with
diagrams, rather than with the whole category or subcategories.
Much of our work will be concerned with diagrams from Top and Set.
A diagram is called <u>commutative</u> if whenever two morphisms are
obtained by composing morphisms in the diagram along different
routes between the same endpoints, the two morphisms are equal.

Example 1

The associative law of composition in any category (cf. §1.2. iii)
is equivalent to the
requirement that the
following diagram is
commutative whenever
the constituent
compositions are
defined.

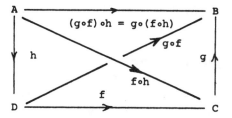

Example 2

The composition rule for identity morphisms in any category
(cf. §1.2. iv) is equivalent to the requirement that the following
diagram is commutative.

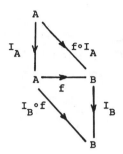

Example 3

The diagram (in Set, Top and Grp)

$$\Delta = \{ \mathbb{R} \overset{a}{\underset{b}{\to}} \mathbb{R} \,|\, a : x \longmapsto 2x, \; b : x \longmapsto 3x \}$$

is plainly not commutative.

Remark

We shall henceforth use only diagrams that are <u>small</u> in the technical sense : the collections of objects and morphisms in a diagram will always belong to some fixed universe of sets. This does not restrict us to finite or even countable diagrams, of course. Diagrams crop up naturally in physics and much theory is devoted to their interpretation. Category theory provides, among other things, an abstract calculus for diagram chasing and in particular it prescribes certain natural objects and morphisms that may be available, by standard construction, for incorporation in given diagrams.

1.7   *Natural transformation*

Given two (covariant) functors F, G from C_1 to C_2 then a <u>natural transformation</u> from F to G is denoted by $\tau : F \overset{\cdot}{\to} G$ and defined to be a map

$$\tau : \text{Objects of } C_1 \longrightarrow \text{Morphisms of } C_2 : A \longmapsto \tau(A)$$

such that every diagram

yields in C_2 the
following commutative
diagram.

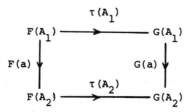

A natural transformation $\tau : F \overset{\bullet}{\to} G$ is called a <u>natural equivalence</u>
or <u>natural isomorphism</u> if

$$\left(\forall \tau(A)\right)\left(\exists \tau(A)^{-1}\right) : \begin{cases} \tau(A) \circ \tau(A)^{-1} & = & I_{G(A)} \\ \tau(A)^{-1} \circ \tau(A) & = & I_{F(A)} \end{cases}$$

in that case F and G are called <u>naturally isomorphic functors</u>.

<u>Example 1</u>

Natural transformations compose to give a natural transformation.
Hence we can obtain $C_2{}^{C_1}$ a <u>functor category</u> whose morphisms are
natural transformations and whose objects are functors from C_1
to C_2.   The objects of $C_1{}^{C_2\mathrm{op}}$ are contravariant functors from
C_2 to C_1 (cf. §1.2 Ex.6).

<u>Example 2</u>

For any functor F there is a natural isomorphism $1_F : F \overset{\bullet}{\to} F$
defined by $1_F(A) = I_{F(A)}$.

<u>Example 3</u>

For finite-dimensional vector spaces we have the most celebrated
natural isomorphism, the <u>taking of double duals</u>.   Here, F is
the identity functor and

$$G : f \begin{array}{ccc} V & & V^{**} \\ \bigg\downarrow & \longrightarrow & \bigg\downarrow f^{**} \\ W & & W^{**} \end{array}$$

where (cf. §1.4. Ex. 1)

$$V^{**} = (V^*)^* = L(V^*;I\!R)$$

and the constituent maps of the functor G are given by

$$\left\{ \begin{array}{l} V \longrightarrow V^{**} : x \longmapsto x^{**} \left\{ \begin{array}{l} x^{**} : V^* \longrightarrow I\!R \\ : \alpha \longmapsto \alpha(x) \end{array} \right. \\ L(V;W) \longrightarrow L(V^{**};W^{**}) : f \longmapsto f^{**} \left\{ \begin{array}{l} f^{**} : V^{**} \longrightarrow W^{**} \\ : a \longmapsto a \circ f^*. \end{array} \right. \end{array} \right.$$

The importance of these maps lies in their freedom from arbitrary choices; for example, they do not depend on what bases are used in the spaces. This freedom is what is referred to as naturality in their construction; it is characteristic of natural isomorphisms. In physical theories, often the term canonical is used with a similar meaning.

1.8 Special morphisms

In a category C a morphism $f : A \longrightarrow B$ is called :

(i)   invertible in C if $\exists g : B \longrightarrow A$ in C with $g \circ f = I_A$ and $f \circ g = I_B$, then we write $A \equiv B$ and say that A and B are isomorphic and f is an isomorphism in C;

(ii)   monic (or a monomorphism) in C if whenever there are

$$k, \ell : C \longrightarrow A \qquad \text{with } f \circ k = f \circ \ell$$

then $k = \ell$;

(iii) epic (or an epimorphism) in C if whenever there are

$$k, \ell : B \longrightarrow D \qquad \text{with } k \circ f = \ell \circ f$$

then $k = \ell$.

Example 1

In Set the monics are the injections (one-to-one non-empty maps)
the epimorphisms are the surjections (onto maps) and the
isomorphisms are the bijections (one-to-one and onto maps.)

Example 2

In Top, the morphisms are continuous maps and isomorphisms are
homeomorphisms.   It is well known that though every continuous
bijection has an inverse map, it need not be continuous (try
mapping an interval round a circle) and hence need not be
invertible in Top, though it is invertible in Set.

Example 3

In the category Rng of rings and ring maps, the inclusion map of
the integers in the rationals is epic and monic but not an
isomorphism.

1.9   *Special Objects*

In a category C an object A is called

(i)   terminal if to each object B in C there is exactly one
        morphism B $\longrightarrow$ A;

(ii)   initial if to each object B in C there is exactly one
        morphism A $\longrightarrow$ B

(iii) null or zero if it is both terminal and initial.

Remark

I_A is the only morphism A $\longrightarrow$ A if A is terminal;   also, any
two terminal objects are isomorphic and any two initial objects are
isomorphic.

Example 1

In Set (and Top), any one point set {x} is a terminal object, and
the empty set is an initial object (the empty map is the unique
morphism from the empty set to any set).

Example 2

In Grp, the singleton group   ({e}, ∘)   is a null object;   similarly
the trivial vector space   {0}   in Vec.

Example 3

The ring of integers $(Z , +, \cdot)$ is an initial object in the
category of rings and ring-maps and the singleton rings like
({0}, +, ·) are terminal objects.

Example 4

The category of fields and field-maps has no initial object and no
terminal object (recall that the smallest field has two elements,
0 and 1, but in a ring we may have 0 = 1).

1.10   *Inverse morphisms*

Given a morphism $f : A \longrightarrow B$ in a category C ,

(i)   a right inverse or section or coretraction of f is a
       morphism $g : B \longrightarrow A$ with $f \circ g = I_B$;

(ii)   a left inverse or retraction for f is a morphism
       $h : B \longrightarrow A$ with $h \circ f = I_A$.

Remark

If f has a section g then f is epic (cf. §1.8 iii).   For suppose
also $k, \ell : B \longrightarrow D$ and $k \circ f = \ell \circ f$;   then
$k \circ f \circ g = k \circ I_B = k = \ell \circ f \circ g = \ell \circ I_B = \ell$.   The assertion
that every set-theoretic surjection has a section is called the
axiom of choice in set theory;   it is equivalent to Zorn's Lemma.
If f has a retraction h then f is monic (cf. §1.8 ii).   For
suppose also $k, \ell : C \longrightarrow A$ and $f \circ k = f \circ \ell$;   then
$h \circ f \circ k = I_A \circ k = k = h \circ f \circ \ell = I_A \circ \ell = \ell$.

Example 1

In Set a morphism is a section if and only if it is injective and
not the empty map from ∅ to a non-empty set;   but in Grp this
condition is necessary though not sufficient.

Example 2

In Top, g : B ⟶ A is a section of f : A ⟶ B if and only
if g is a continuous injection, continuously invertible on
g(B) ⊆ A.

Example 3

In Set a morphism is a retraction if and only if it is surjective.

Example 4

A tangent vector field on a smooth manifold M is a section
V : M ⟶ TM of the canonical surjection Π : TM ⟶ M.   This
simply means that Π∘V = I$_M$ and so we know that V attaches to
each point x∈M a vector V(x) from T$_x$M, the tangent vector
space to M at x.

2.   STRUCTURES ON CATEGORIES

We have in the previous section the basic terms in a language that
exploits commonly occurring patterns in mathematical constructions.
In applications to physical theories these same distinctive
patterns have significant interpretations and often the interplay
of different branches of mathematics can be viewed through the
roles of functors.   Frequently, the objects (e.g., geometrical
or configuration spaces) in physical theories are derived from
more primitive objects (usually potentially observable interactions)
and a similar thing happens in mathematics.   It turns out that
many of these processes of derivation are examples of the taking
of limits of diagrams in a category.   Closely related to natural
constructions being carried from one category to another is the
concept of adjoint functors.   In this section we shall pursue
these two ideas.

2.1   Limits of diagrams

Denote by Δ the diagram (cf. §1.6)

$$\{a_{ij}^k : A_i \longrightarrow A_j \,|\, k \in J_m;\ i,\ j \in J_o\}$$

in a category C, for some indexing sets J_m, J_o .   We define as follows:

(i) A <u>left limit</u> (if it
     exists) of Δ is an
     object L and morphisms

     $\{f_i : L \longrightarrow A_i | i \epsilon J_o\}$

     in C , commuting with
     the a_{ij}^k and such that
     if

     $\{g_i : K \longrightarrow A_i | i \epsilon J_o\}$

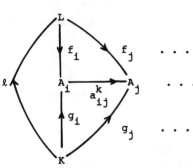

in C also commutes with

the a_{ij}^k then there exists a <u>unique</u> morphism

$\ell : K \longrightarrow L$ such that $(\forall i \epsilon J_o)$ $g_i = f_i \circ \ell$. If L is

such a left limit object of Δ then we sometimes write

(loosely) $L = \varprojlim \Delta$, the existence of the morphisms,

and ℓ with the universal property, being understood. We

can speak of <u>the</u> left limit because, if it exists, it is

determined up to isomorphism in C .

<u>Remark</u>

Left limits are sometimes called just <u>limits</u> or <u>projective limits</u>
or <u>left roots</u>, or <u>inverse limits</u>.

(ii) A <u>right limit</u> (if it
      exists) of Δ is an
      object R and
      morphisms

      $\{f_i : A_i \longrightarrow R | i \epsilon J_o\}$

      in C , commuting with
      the a_{ij}^k and such that
      if

      $\{g_i : A_i \longrightarrow K | i \epsilon J_o\}$

      in C also commutes with the
      a_{ij}^k then there exists a
      <u>unique</u> morphism $r : R \longrightarrow K$
      such that $(\forall i \epsilon J_o)$ $g_i = r \circ f_i$.

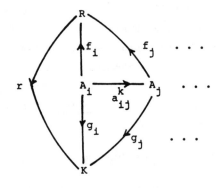

Then, as before, such a limit is determined up to isomorphism in
C , and we write here $\lim_{\rightarrow} \Delta$ = R.

Remark

Right limits are sometimes called <u>colimits</u>, <u>inductive limits</u> or
<u>right roots</u> or <u>direct limits</u>.

Example 1

Consider in Set any diagram like $\{A_1 \xleftarrow{\ a_1\ } A_0 \xrightarrow{\ a_2\ } A_2\}$.
This always has left limit object

$$L = A_0 \text{ with } f_i = a_i \text{ for } i = 1, 2 \text{ and } f_0 = I_{A_0}.$$

For suppose we have

$\{g_i : K \longrightarrow A_i | i=0,1,2\}$

commuting with the
a_i, then
$g_0 : K \longrightarrow A_0$ is the
required unique map with

$g_i = f_i \circ g_0.$

(Observe that if we omit A_1 and a_1 from our diagram the left limit
is unchanged.)

Example 2

Consider, in Set, any diagram like $\{A_1 \xrightarrow{\ a_1\ } A_0 \xleftarrow{\ a_2\ } A_2\}$.
This always has right limit object $R = A_0$ with $f_i = a_i$ for
$i = 1, 2$ and $f_0 = I_{A_0}$.

For suppose we have

$\{g_i : A_i \longrightarrow K | i = 0,1,2\}$

commuting with the a_i,
then $g_0 : A_0 \longrightarrow K$ is
the required unique map
with $g_i = g \circ f.$ (If we
omit A_2, a_2 from our
diagram the right limit is unchanged.)

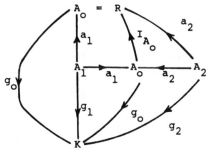

Example 3

Consider in Set the diagram $\{A_1, A_2\}$, consisting of two sets and no morphisms.
The left limit is $A_1 \times A_2$, the <u>Cartesian product</u> consisting of all ordered pairs of elements from A_1 and A_2. ($A_1 \times A_2 = \emptyset$ if at least one of A_1, A_2 is $\emptyset$.) In this case each f_i is projection onto the i-th component and the map

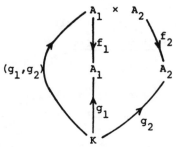

$$(g_1, g_2) : K \longrightarrow A_1 \times A_2 : x \longmapsto (g_1(x), g_2(x))$$

is evidently unique.

Example 4

Consider again in Set the diagram $\{A_1, A_2\}$, consisting of two sets and no morphisms. It has right limit $A_1 \amalg A_2$, the <u>disjoint union</u> of A_1 and A_2, which is the union of disjoint copies of A_1 and A_2.
There are obvious injections
$f_i : A_i \longrightarrow A_1 \amalg A_2$ and the required map
$r : A_1 \amalg A_2 \longrightarrow K$ is uniquely determined by
$r : x \longmapsto g_i(x)$ if $x \in \overset{\leftarrow}{f_i} A_i$.

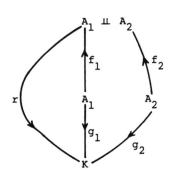

Remark

We shall encounter these four examples again to find the other limits in Examples 1, 2 but first we generalize 3, 4 to arbitrary collections of sets.

2.2   *Products and coproducts*

Let J be an index set and consider a diagram $\Delta_o = \{A_i | i \in J\}$ with
no morphisms, in a category C .

(i)   If it exists, the left limit object $\underleftarrow{\lim} \Delta_o$ is called the
       product in C of the objects $\{A_i | i \in J\}$; it is denoted
       by $\prod\limits_{i \in J} A_i$ and the associated maps $\pi_j : \prod\limits_{i \in J} A_i \longrightarrow A_j$ are
       called projections.

(ii)   If it exists, the right limit object $\underrightarrow{\lim} \Delta_o$ is called the
       coproduct in C of the objects $\{A_i | i \in J\}$; it is denoted
       by $\coprod\limits_{i \in J} A_i$ and the associated maps $\mu_j : A_j \longrightarrow \coprod\limits_{i \in J} A_i$ are
       called injections.

Example 1

As we expect from §2.1 Ex. 3, in Set the product of objects
$\{A_i | i \in J\}$ does exist, it is their Cartesian product with projections
$\pi_j : \prod\limits_{i \in J} A_i \longrightarrow A_j : (x_i)_{i \in J} \longmapsto x_j.$

Example 2

The product of objects $\{(A_i, \circ) | i \in J\}$ in Grp is the direct product
group $(\prod\limits_{i \in J} A_i, \circ)$ where $\prod\limits_{i \in J} A_i$ is the set product and
composition is

$$\left(a_i\right)_{i \in J} \circ \left(b_i\right)_{i \in J} = \left(a_i \circ b_i\right)_{i \in J} .$$

Example 3

The product of topological spaces $\{(A_i, T_i) | i \in J\}$ in Top
(cf. III §1 below) again uses the set product and has the smallest
topology (cf. III §2 below) that makes the projections continuous.
So $\prod\limits_{i \in J} A_i$ has as open sets only those U such that for some
$U_i \in T_i$ (i.e., U_i is open in A_i) $\pi_i U = U_i$. Hence U must
be of the form $\bigcap \overleftarrow{\pi_i} U_i$, for some open $U_i \subseteq A_i$.

Example 4

Further to §2.1 Ex. 4, in Set the coproduct of objects $\{A_i | i \in J\}$
does exist, it is their disjoint union, $\coprod\limits_{i \in J} A_i$ with injections

$$f_j : A_j \longrightarrow \coprod_{i \in J} A_i : x \longmapsto x \ ,$$

and any collection of maps

$$\{g_j : A_j \longrightarrow K | j \in J_o\}$$

determines a unique map

$$r : \coprod_{i \in J} A_i \longrightarrow K : (x \in A_j) \longmapsto g_j(x) \ .$$

Example 5

Coproduct carries over to Grp, to give the <u>free product group</u> of a collection of groups, using the disjoint operations on the disjoint union. In Top we give $\coprod_{i \in J} A_i$ the <u>largest</u> topology that gives continuity to the injections, so $U \subseteq \coprod_{i \in J} A_i$ is open only if for <u>all</u> $i \in J$ we can find open $U_i \subseteq A_i$ with $f_i^{\leftarrow} U = U_i$. The topological space so formed is called the <u>topological disjoint sum</u> of the spaces involved.

Example 6

In the case that J is a singleton, then the diagram consists of one object and it is its own left and right limit. To see this, omit A_1, A_2 and a_1, a_2 from the diagram in §2.1 Ex.1.

2.3   *Pullback and pushout*

(i)   In any category, given a diagram

$$A_1 \xrightarrow{\ a_1\ } A_o \xleftarrow{\ a_2\ } A_2$$

then, if it exists, its left limit is called the <u>pullback</u> (or <u>fibred product</u> of A_1 and A_2) over A_o and the completed diagram so formed is called a <u>pullback square</u>; it is <u>commutative</u>.

Remark

Either of $a_1 \circ f_1$, $a_2 \circ f_2$ will

serve as $f_o : L \longrightarrow A_o$ (cf. §2.1 i).

<u>Example 1</u>

In Set we construct the pullback as follows:

$$L = \prod_{i=1,2} X_i \quad \text{with} \quad X_1 \subseteq A_1, \; X_2 \subseteq A_2$$

given by

$$X_1 = \{x \in A_1 \,|\, a_1(x) \in a_2 A_2\}$$
$$X_2 = \{y \in A_2 \,|\, a_2(y) \in a_1 A_1\} \;.$$

From the definition of left limit (cf. §2.1 i) we know that given any

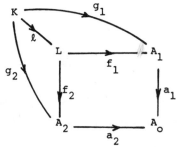

$$A_1 \xleftarrow{\;g_1\;} K \xrightarrow{\;g_2\;} A_2$$

commuting with a_1, a_2 , there is a unique morphism $\ell : K \longrightarrow L$ with $g_1 = f_1 \circ \ell$ and $g_2 = f_2 \circ \ell$. From §2.1 Ex.2 we know that the original diagram always has a right limit, namely A_o.

(ii) There is a dual to (i) : given a diagram

$$A_1 \xleftarrow{\;a_1\;} A_o \xrightarrow{\;a_2\;} A_2$$

in any category then, if it exists, its right limit is called the <u>pushout</u> (or <u>fibred coproduct</u> or <u>fibred sum</u> of A_1 and A_2) over A_o and the completed diagram so formed is called a <u>pushout square</u> ; it is <u>commutative</u>.

<u>Remark</u>

By commutativity we have

$$f_1 \circ a_1 = f_2 \circ a_2 = f_o : A_o \longrightarrow R \;.$$

Example 2

In Set we find R from an equivalence relation on $\coprod_{k=1,2} A_k$

defined by : $x \sim y$ if any of the following hold

(x) $x = y$

(β) $a_1(x) = y$ or $a_1(y) = x$

(f) $a_2(x) = y$ or $a_2(y) = x$.

Then   R is the set of equivalence classes

$$R = \{[x]_\sim \mid x \in \coprod_{k=1,2} A_k\}$$

where

$$[x]_\sim = \{y \in \coprod_{k=1,2} A_k \mid x \sim y\}$$

and the associated morphisms are (cf. §2.1 ii)

$f_i : A_i \longrightarrow R : x \longmapsto [x]_\sim$ sending elements to their
equivalence classes.   These morphisms are well-defined because we
always look on the A_i as disjoint subsets of $\coprod_{k=1,2} A_k$.

It follows that given any

$$A_1 \xrightarrow{g_1} K \xleftarrow{g_2} A_2$$

commuting with a_1, a_2,
there is a unique morphism
$r : R \longrightarrow K$ such that
$g_1 = r \circ a_1$ and $g_2 = r \circ a_2$.
From §2.1 Ex. 1 we know
that the original diagram
always has A_o as left
limit.

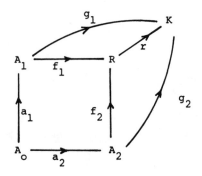

Example 3

In Set, given subsets A_1, A_2 of a set A_o we have inclusion
maps a_1, a_2 :

$$A_1 \overset{a_1}{\hookrightarrow} A_o \overset{a_2}{\hookleftarrow} A_2$$

and the pullback object is
the intersection $A_1 \cap A_2$.

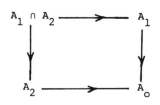

Example 4

The same construction as in
Example 3 serves in Grp for
subgroups A_1, A_2 of a
group A_o (when necessarily the inclusion maps are group
homomorphisms) and in Top for subsets (when necessarily the
subspace topologies make the inclusion maps continuous).

Example 5

A typical situation that arises in bundle theory is the diagram

$$X \xrightarrow{\Pi_X} M \xleftarrow{\Pi_Y} Y$$

in Man, the category of smooth manifolds and smooth maps among
them.   Here X and Y are bundles over M with canonical
projections Π_X, Π_Y onto M.
We obtain a pullback square with
$L = \{(x,y) \in X \times Y \mid \Pi_X(x) = \Pi_Y(y)\}$
and the submanifold structure
induced by inclusion in the
product manifold X×Y.   The
smooth maps π_1, π_2 are just
the restrictions of the usual

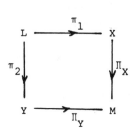

projections from a Cartesian product set (cf. §2.2 Ex. 1).

2.4   *Equalizer and coequalizer*

In any category, given a diagram

$$\Delta : \quad A_1 \mathrel{\substack{\xrightarrow{a_1} \\ \xrightarrow[a_2]{}}} A_2$$

(i)   the <u>equalizer</u> is the left limit   and

(ii)   the <u>coequalizer</u> is the right limit, when they exist.

Example 1

In Set the equalizer object is

$$L = \{x \epsilon A_1 \mid a_1(x) = a_2(x)\}$$

and the associated morphisms

$f_i : L \longrightarrow A_i$ are
inclusion in A_1 and its
composite with the restriction
of a_1 (or a_2) to $L \subseteq A_1$.

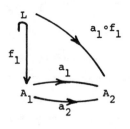

Example 2

The construction in the previous example carries over to Grp, Vec and Top by viewing L as a subobject of the object A_1 .

Example 3

In Set the coequalizer object
is $R = A_2/\sim$ where $\sim$ is the
smallest equivalence relation
on A_2 that contains

$$\{(a_1(x), a_2(x)) \mid x \epsilon A_1\}$$

and the associated morphisms are

$$f_2 : A_2 \longrightarrow R : x \longmapsto [x]_\sim, \quad f_1 = f_2 \circ a_1 .$$

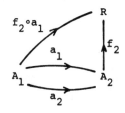

Example 4

In Top we use the coequalizer from Set with the quotient topology: $U \subseteq A_2/\sim$ is open if $\overleftarrow{f_2}U$ is open in A_2 . We also say that the topology for $A_2/\sim$ is <u>coinduced</u> by f_2 from A_2 , (cf. III §1).

2.5 *Complete category*

A category C is called

(i)   <u>left complete</u> if every diagram has a left limit in C ,

(ii)   <u>right complete</u> if every diagram has a right limit in C ,

(iii) <u>complete</u> if both left complete and right complete,

(iv)   <u>finitely</u> left complete/right complete/complete if in (i), (ii), (iii) respectively we use only _finite_ diagrams.

Warning

Some authors use complete and cocomplete instead of left complete
and right complete.

Example 1

Some commonly occurring complete categories are : Set; Grp; Top;
R-Vec; Ab (Abelian groups); Rng (rings); R-Mod (modules over a
ring R); Ab_{TF} (torsion free abelian groups); Cat; products of
complete categories.

Example 2

The categories of finite sets and finite topological spaces are
finitely complete but neither is complete.   The category Met of
metrizable topological spaces and continuous maps among them is
countably left complete (all diagrams with countably many
morphisms have left limits) but not complete.

Example 3

In a small category C (so the collections of objects and
morphisms are sets) :

C is left complete <==> C is right complete.

Example 4

The category of finite groups is finitely left complete but not
finitely right complete.

Example 5

The category of non-empty sets is neither finitely left nor
finitely right complete.

Example 6

The category of fields and field maps is neither finitely left nor
finitely right complete.

Example 7

A poset (partially ordered set) is a category and it is finitely
complete if and only if it is a lattice with a smallest member and

a largest member.

Example 8

The actual construction of the limits in Top and Grp relies heavily
on that in Set.   Also, Cartesian products and disjoint unions are
basic constructions for any limits in Set (cf. §2.1 Exs. 3, 4 and
§2.2 Exs. 1, 4).   We can say that for any diagram

$$\Delta = \{a_{ij}^{k} : A_i \longrightarrow A_j \mid k \in J_m, \; i,j \in J_o\} \quad \text{in Set,}$$

(i) $\underset{\leftarrow}{\text{Lim}}\Delta$ is a subset $\prod_{i \in J_o} X_i$ of $\prod_{i \in J_o} A_i$ with $a_{ij}^{k} X_i = X_j$ for

all $i,j \in J_o$, $k \in J_m$ and the restricted projections ;

(ii) $\underset{\rightarrow}{\text{Lim}}\Delta$ is the set of equivalence classes, $\{[x]_\sim \mid x \in \coprod_{i \in J_o} A_i\}$,

of the smallest equivalence relation generated by

$$x \sim y \quad \text{if} \quad (\exists a_{ij}^{k}) : a_{ij}^{k}(x) = y \; ,$$

and the limit morphisms are the maps onto classes.   (For a
discussion and proof see Arbib and Manes [1] p.47.)

Example 9

A category is left complete if and only if it admits products and
equalizers;   it is right complete if and only if it admits
coproducts and coequalizers.   (For proof see Higgins [47] p.54.)

Example 10

All categories are n-complete for n = 1,2,3.   By this we mean
that in any category there are left and right limits for all
diagrams like

A   (cf. §2.2 Ex. 6)

$A \xrightarrow{\;a\;} B$ (cf. §2.1 Exs. 1 and 2, final observations)

$A \xrightarrow{\;a\;} B$

Example 11

For any category C the following are equivalent (cf. Herrlich and Strecker [46] p.158 for more) :

(i) C is finitely left complete;

(ii) C has (all) pullbacks **and a** terminal object;

(iii) C has (all) finite products and (all) pullbacks;

(iv) C has (all) finite products and equalizers.

This proposition, like others in category theory, has a dual
form by changing terms as follows:

left complete	$\longrightarrow$	right complete
pullback	$\longrightarrow$	pushout
product	$\longrightarrow$	coproduct
equalizer	$\longrightarrow$	coequalizer

For more discussion of this duality principle see MacLane [67]
p.31 et seq. also Herrlich and Strecker [46] throughout.

2.6   Limit preserving functors

Given a diagram Δ in a category C_1 and a functor F from C_1
to C_2 we obtain a diagram in C_2 which we can denote by $F(\Delta)$.
Then we say :

(i)   F preserves left limits if, for arbitrary diagrams Δ ,

$$\left(\exists \varprojlim \Delta\right) \Rightarrow \left(\exists \varprojlim F(\Delta)\right) = \left(F \varprojlim \Delta\right) ;$$

(ii)   F preserves right limits if, for arbitrary diagrams Δ ,

$$\left(\exists \varinjlim \Delta\right) \Rightarrow \left(\exists \varinjlim F(\Delta)\right) = \left(F \varinjlim \Delta\right) ;$$

(iii) F preserves limits or is continuous if it perserves both
       left and right limits.

Example 1

The forgetful functors from Grp and RMod to Set do not preserve
right limits.

Example 2

The functor from Set to Ab, constructing free Abelian groups on
sets does not preserve limits.

Example 3

The forgetful functor from Top to Set preserves limits.

Example 4

The inclusion functor Ab ⟶ Grp preserves left limits but not
right limits.

Example 5

The forgetful functor from the category of finite Abelian groups
to Set preserves finite left limits but not arbitrary left limits.

2.7   *Functors of two variables*

We need to prepare for a definition in the next section by
bringing together the notions of opposite category and product
category (cf. §1.2 Exs. 6, 7).

Given three categories C_1, C_2 and C, a functor

$$F : C_1^{op} \times C_2 \longrightarrow C$$

is called a <u>functor of two variables</u>, <u>contravariant in the first</u>
and <u>covariant in the second</u>.

Plainly the terminology extends easily to more variables, and
for the physicist it is reminiscent of that in tensor analysis
(cf. Dodson and Poston [28]).

Example 1

The illustration we need below is the standard case, for any
category C,

$$H_C : C^{op} \times C \longrightarrow Set$$

and

$$H_{Cobj} : (A,B) \longmapsto \{morphisms\ A \longrightarrow B\}$$

$$H_{Cmor} : \quad
\begin{array}{cc}
A_1 & B_1 \\
\uparrow{\scriptstyle f} & \downarrow{\scriptstyle g} \\
A_2 & B_2
\end{array}
\quad \longmapsto \quad
\begin{array}{c}
\{morphisms\ A_1 \xrightarrow{\ h\ } B_1\} \\
\downarrow \\
\{morphisms\ A_2 \xrightarrow{\ g \circ h \circ f\ } B_2\}
\end{array}$$

So what we are doing is
to make this square
diagram commute :

2.8   *Adjoint functors*

Certain functors crop up in pairs, one going each way, and they
are important because they preserve limits (cf. §2.6) and because
they characterise certain general constructions (cf. Higgins [47]
p.15 et seq.).

Given two (covariant) functors

$$F : C_1 \longrightarrow C_2 \quad \text{and} \quad G : C_2 \longrightarrow C_1$$

we can by the construction in §2.6 obtain two functors

$$\tilde{F}, \ \tilde{G} : C_1^{op} \times C_2 \longrightarrow \text{Set}$$

by making compositions with H_{C_1}, H_{C_2} :

$$
\begin{array}{ccc}
C_1^{op} \times C_2 & \xrightarrow{\ (I_1^{op},G)\ } & C_1^{op} \times C_1 \\
(F,I_2) \downarrow & \text{Need not be} & \downarrow H_{C_1} \\
 & \text{commutative!} & \\
C_2^{op} \times C_2 & \xrightarrow[\ H_{C_2}\ ]{} & \text{Set}
\end{array}
\qquad
\left\{
\begin{array}{l}
\tilde{F} = H_{C_2} \circ (F,I_2) \\[2ex]
\tilde{G} = H_{C_1} \circ (I_1^{op},G)
\end{array}
\right.
$$

In general this category diagram is not commutative.   The
next best thing is for the two composite functors $\tilde{F}$ and $\tilde{G}$ to
be naturally equivalent (cf. §1.7); in that case we say that
(F,G) is an <u>adjoint pair</u> of functors.   We also say that F <u>is
left adjoint to</u> G,   or F <u>is a left adjoint</u> and F <u>has a right
adjoint</u>.   This relation between functors is not symmetric
(cf. Ex. 5 below).

Remark

The intricacy of the definition somewhat obscures a fairly simple
intuitive idea.   Namely, that if (F,G) is an adjoint pair then:

(i)   in C_1 ,   there are morphisms from objects in C_1 to
      objects in $G_{obj}C_2$ (e.g. $A \xrightarrow{f} G_{obj}B$) ;

(ii) in C_2 ,   there are morphisms from objects in $F_{obj}C_1$ to
      objects in C_2 (e.g. $F_{obj}A \xrightarrow{f^\dagger} B$) ;

(iii) there is a correspondence $f \longmapsto f^\dagger$ such as to make both
      of the following diagrams commute whenever one of them
      commutes.

                 in C_1 in C_2

Theorem   (The adjoint functor theorem)

Given a category C_1 whose collection of morphisms between its
objects is a set and admitting limits for all diagrams, then a
functor $G : C_1 \longrightarrow C_2$ has a left adjoint if and only if G
preserves limits of all diagrams and satisfies :

For each object B in C_2 there is a set J and morphisms

$$\{f_i : B \longrightarrow GA_i \,|\, i \in J\}$$

such that every morphism $g : B \longrightarrow GA$ can be written as a
composite $g = (Gh) \circ f_i$ for some $i \in J$ and some $h : A_i \longrightarrow A$.

Proof

MacLane [67] p.117.   □

Remark

This theorem is important because it suggests the construction
process needed for the required left adjoint.   We collect below
the left adjoints of some common forgetful functors.   For
constructions and more examples see Herrlich and Strecker [46].

| Categories | | Adjoint functors | |
C_1	C_2	F left adjoint	G right adjoint
Grp	Set	free group	forgetful
Ab	Grp	factor by commutator group	inclusion
Top	Set	discrete space	forgetful
Complete metric spaces	Metric spaces	completion	inclusion
Complete uniform spaces	Uniform spaces	uniform completion	inclusion
Locally convex Top vec	Top vec	locally convex modification	inclusion
Compact Haus	Set	Stone – Čech discrete compactification	forgetful

Example 1

If a functor F has a right (or left) adjoint then that adjoint
is uniquely (up to natural equivalence) determined by F.   (For a
proof see Higgins [47] p.17 et seq.)

Example 2

(Cf. Strooker [97] p.25)   Let E be a fixed set and let F be
the functor from Set to Set given by the maps

$$F_{obj} : A \longrightarrow A \times E$$

$$F_{mor} : \quad \begin{array}{c} A_1 \\ \downarrow f \\ A_2 \end{array} \quad \longrightarrow \quad \begin{array}{c} A_1 \times E \\ \downarrow (f, I_E) \\ A_1 \times E \end{array}$$

Also let G be the functor from Set to Set given by

$$G_{obj} : A \longrightarrow \{E \longrightarrow A\} = \{\text{all maps from } E \text{ to } A\}$$

$$G_{mor} : \quad \begin{array}{c} A_1 \\ \downarrow f \\ A_2 \end{array} \quad \longrightarrow \quad \begin{array}{c} \{E \xrightarrow{h} A_1\} \\ \downarrow \\ \{E \xrightarrow{f \circ h} A_2\} \end{array}$$

Then (F,G) are adjoint functors.

Example 3

(This is the important theorem concerning adjoint functors and limits, cf. Higgins [47] p.52 for proof.)

Let $F : C_1 \longrightarrow C_2$ and $G : C_2 \longrightarrow C_1$ be functors such that (F,G) is an adjoint pair.   Then F preserves right limits and G preserves left limits.

Example 4

The forgetful functor

$$G : \text{Top} \longrightarrow \text{Set}$$

has as left adjoint, the discrete topologising functor

$$F : \text{Set} \longrightarrow \text{Top} : A \longmapsto (A, \text{Sub } A)$$

so (F,G) is an adjoint pair.

   Also, G has a right adjoint, the indiscrete topologising functor

$$H : \text{Set} \longrightarrow \text{Top} : A \longmapsto (A, \{A, \emptyset\}),$$

but H has no right adjoint (cf. MacLane [67] p.131).

Example 5

The category Haus of all Hausdorff spaces is a full subcategory (cf. §1.5) of Top, and it is complete.
The inclusion functor

$$G_1 : \text{Haus} \longrightarrow \text{Top}$$

and the forgetful functor

G_2 : Haus $\longrightarrow$ Set

both have left adjoints, but neither has a right adjoint.
For a proof see MacLane [67] p.131-2.

We conclude this chapter by drawing attention to the extensive
bibliography on category theory provided in Herrlich and Strecker
[46] pp. 332-380 and also to the exceptionally detailed index in
that book.   Both will be found very useful by those wishing to
pursue further the introductory treatment we have given above.

III Existence of limiting topologies

We can think of topology as the abstract study of continuity for
maps between sets.   In a sense, having a topology is the least
extra structure that a set can be given in order for the notion
of continuity to have meaning for maps to and (or) from the set.
There are many good introductions to topology : Jameson [54] and
Maddox [68] slanted towards normed spaces, Porteous [78] slanted
towards topological algebra and Wall [100] slanted towards
algebraic topology.   As a general reference work Császár [20] is
invaluable and we shall use it frequently.   The goal of this
chapter is to prove the existence of certain limiting topologies
by exploiting a partial ordering of topologies on a given set.
That will be done in the second section.   First we collect for
our later convenience some of the basic terminology and elementary
constructions.   We omit proofs and examples from the first section
since they are readily available elsewhere, but mainly we give
full details in the second section.   There we prove that Top is a
complete category and we look at some of its limiting structures.
Recent papers on categorical topology can be found in the volume
edited by Binz and Herrlich [4].

1.   BASIC TOPOLOGY

We shall mainly work from the viewpoint of open sets rather than
from neighbourhoods (cf. Jameson [54], Császár [20]).   For a more
leisurely motivation of the definitions see Dodson and Poston [28]
where the goal is to provide for continuity and differentiability
on vector spaces as a preliminary to the study of manifolds and

applications in relativity.   For testing your intuition in
topological arguments the study by Steen and Seebach [95] of
counterexamples can be helpful!

1.1   *Topological space and continuous map*

A topological space is a pair (X,T) where X is a set and T
is a collection of subsets of X satisfying :

(i) $\emptyset$, $X \in T$,

(ii)   arbitrary unions of sets in T are in T ,

(iii) finite intersections of sets in T are in T .

The collection T is called a topology for X.

A continuous map from a topological space (X_1, T_1) to a
topological space (X_2, T_2) is a (set-theoretic) map

$f : X_1 \longrightarrow X_2$ such that $f^{\leftarrow}B \in T_1$ $(\forall B \in T_2)$.

The elements of a topology T for X are called open sets
of (X,T), or T-open sets of X, and their complements in X
are called closed sets of (X,T) , or T-closed sets of X.

Remark

Continuity of maps is preserved by map composition but not by
inversion of maps.   A continuous bijection with a continuous
inverse is called a homeomorphism (i.e. a topological isomorphism)
and when one exists between two topological spaces we call them
homeomorphic.   Being homeomorphic is an equivalence relation $\approx$
on topological spaces.   A surjective map is called open (respectively
closed) if the image of every open (closed) set is open (closed).
It follows that a bijection is a homeomorphism if and only if it
is continuous and open.

Top is the category of all topological spaces with continuous
maps among them (cf. II §1.2), we shall write these morphisms in
the form

$f : (X_1, T_1) \longrightarrow (X_2, T_2)$.

Recall that a pseudo-metric or semimetric on a set X is a
map

$$\rho : X \times X \longrightarrow I\!R$$

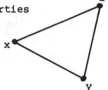

satisfying for all x, y, z∈X the properties

(i) $\rho(x,x)$ = 0

(ii) $\rho(x,y)$ = $\rho(y,x)$

(iii) $\rho(x,z)$ ⩽ $\rho(x,y)$ + $\rho(y,z)$.

A set X with such a map is called a <u>pseudo-</u> or <u>semimetric</u> space.
We note that by putting x = y = z in (iii) we deduce that ρ is
non-negative.   If ρ is in fact <u>positive definite</u>, that is to
say it satisfies also

(iv) $\rho(x,y)$ = 0 => x = y

then we call it a <u>distance function</u> or <u>metric</u> for X.   Met is the
category of small metric spaces and continuous maps.

    This should not be confused with the use of the word 'metric'
in physics literature to mean a <u>metric tensor field</u> on a manifold
(cf. [28]).

<u>Example</u>

Every semimetric space (X,ρ) is a topological space (X, T_ρ) .
Here, any $A \subseteq X$ is a T_ρ -open set if

    $(\forall x \in A)(\exists$ real r>0) : $S(x,r) \subset A$

where

    $S(x,r)$ = $\{y \in X | \rho(x,y) < r\}$

is the <u>open ball of radius r</u> in (X,ρ).

1.2   <u>Neighbourhood, base and sub base</u>

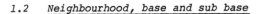

An (open) <u>neighbourhood</u> of any x∈X, when (X, T) is a topological
space, is any subset of X containing some A∈T with x∈A .

    A <u>base</u> for a topology T on X is a collection B of
subsets of X satisfying

(i) $B \subseteq T$

(ii)   every non-empty T-open set is expressible as a union of
       sets from B .

Example

In any metric space (X,ρ) a base for T_ρ is

$$B = \{S(x,r) \mid x \in X, \ r > 0\}$$

the set of all open balls of positive radius in X.

A <u>sub base</u> for a topology T on X is a collection B_s of subsets of X if the set of all finite intersections of sets from B_s form a base for T . (Recall that the empty set is the smallest finite set, it has no elements.)

Remark

A <u>cover</u> of a set X is a collection C of subsets of X with $X \subseteq \bigcup_{A \in C} A$. Hence any cover can be used as a sub base to generate a topology.

A map f from (X_1, T_1) to (X_2, T_2) is called <u>continuous</u> at $x \in X$ if the inverse image of every neighbourhood of f(x) contains a neighbourhood of x. We say that f <u>is continuous</u> if it is continuous at all points of its domain.

1.3   *Separation axioms*

A topological space (X,T) is called :

(i) T_0 if whenever x,y are distinct points in X <u>at least one</u> of them possesses a neighbourhood not containing the other point;

(ii) T_1 if whenever x,y are distinct points in X <u>each of them</u> possesses a neighbourhood not containing the other point;

(iii) T_2 (or Hausdorff) if whenever x,y are distinct points in X they possess neighbourhoods that are mutually disjoint.

It follows that a space is

$T_1 \iff$ every point is a closed set

$T_2 \iff$ every point is the intersection of its closed neighbourhoods.

Example

Metric spaces are always T_2 , and a semimetric space (X,ρ) is T_0 if and only if ρ is a metric.

1.4   *Closure, boundary, compactness, local compactness*

Let (X,T) be a non-empty topological space.

A <u>limit point</u> of a set A⊆X is any p∈X such that:

$$p \in N \in T \implies N \backslash \{p\} \cap A \neq \emptyset \ .$$

The <u>closure</u> A^c of a set A⊆X is the union of A and all of
its limit points;   equivalently A^c is the smallest closed set
containing A .

The <u>interior</u> int A of a set A⊆X is the largest open set
contained in A.

The <u>boundary</u> $\overset{\bullet}{A}$ of a set A⊆X is A^c\int A .

The space (X,T) is <u>compact</u> if every open cover contains a
finite subcover (cf. §1.2), and <u>locally compact</u> if every point has
a compact neighbourhood.   We recall that the continuous image of
a compact set is compact.

1.5   *Connectedness, denseness, separability*

A topological space (X,T) is called <u>connected</u> if it is not the
union of two disjoint non-empty open sets, or equivalently : if
the only sets that are both open and closed are ∅ and X.   A
<u>component</u> of a topological space is a maximal connected subset.
Connectedness is preserved by continuous maps;   components are
closed sets.

A subset A of the space (X,T) is called <u>dense</u> if $A^c = X$,
or equivalently if any neighbourhood of any point of X intersects
A .   The space (X,T) is called <u>separable</u> if it has a countable
dense subset.   Since the rationals are countable and dense in the
reals it follows that $\mathbb{R}^n$ is separable for each finite n.

The space (X,T) is called

<u>first countable</u> if each point has a countable neighbourhood base

<u>second countable</u> if T has a countable base.

It follows that second countable implies first countable and
separable, and Lindelöf's Theorem shows that every open cover of a
second countable space has a countable subcovering.

1.6   *Paracompactness and partitions of unity*

In a topological space   (X,T)   a collection   K   of subsets of   X
is called <u>locally finite</u> if every point of   X   has a neighbourhood
which intersects only a finite number of the sets from   K.   A
collection   K   is called a <u>refinement</u> of a collection   L   if

$$(\forall A \epsilon K)(\exists B \epsilon L) : A \subseteq B$$

   The space   (X,T)   is called <u>paracompact</u> if every open cover
of   X   has a locally finite refinement which is also an open cover
of   X.

   A non-empty family Φ of continuous real functions on   (X,T)
is called <u>locally finite</u> if every point of   X   has a neighbourhood
on which all but finitely many of the functions in Φ take the
value zero.   For such a family, their sum function

$$\sum_{f \epsilon \Phi} f : X \longrightarrow R$$

is continuous and we call Φ a <u>partition of unity</u> if

$$f \epsilon \Phi, \ x \epsilon X \implies f(x) \geqslant 0, \ \sum_{f \epsilon \Phi} f(x) \ = \ 1 \ .$$

Such a partition of unity is called <u>subordinate</u> to a collection   K
of subsets of   X   if

$$(\forall f \epsilon \Phi)(\exists A \epsilon K) : f(x) \ = \ 0 \ \ \forall x \notin A \ .$$

It is easily shown that if in   (X,T)   for every open cover   K   there
is a partition of unity subordinate to   K ,   then   (X,T)   is
paracompact.

   Other sufficient conditions for paracompactness are, for
example, compactness or pseudo-metrizability.   In the case of a
(Hausdorff) <u>smooth manifold</u> (cf. Ch. IV), paracompactness is
sufficient to ensure the existence of a (smooth) partition of unity
which in turn leads to a Riemannian structure and hence metrization.

1.7   *Homotopy, contractibility, covering space*

Given two spaces   X,Y   and two continuous maps f_o, f_1 : X \longrightarrow Y
we say that f_o <u>is homotopic to</u> f_1 if there exists a continuous
map (a <u>homotopy</u>)

$$F : X \times [0,1] \longrightarrow Y$$

satisfying

$$(\forall x \in X) F(x,0) = f_o(x), \quad F(x,1) = f_1(x) .$$

Being homotopic is an equivalence relation $\sim$.

A space X is called <u>contractible</u> if the identity I_X is homotopic to some constant map, $c : X \longrightarrow \{x_o\}$, for some $x_o \in X$.

A space X is called <u>simply connected</u> if it satisfies:

(i) $(\forall x,y \in X)$ $\exists$ continuous curve $c : [0,1] \longrightarrow X$ with
       $c(0) = x$, $c(1) = y$ (<u>arcwise connectedness</u>) ;

(ii) $(\forall x \in X)$ $(\forall$ open neighbourhoods N of $x)$ $(\exists$ open U with
       $x \in U \subset N$ and U arcwise connected); (<u>local arcwise</u>
       <u>connectedness</u>) ;

(iii) every closed curve in X is homotopic to a constant curve.

A space $\tilde{X}$ is called a <u>covering space</u> of X if $\tilde{X}$ satisfies (i) and (ii) and there is a continuous surjection $p : \tilde{X} \longrightarrow X$ such that $(\forall x \in X)$ $(\exists$ open neighbourhood N of $x)$ with $p^{\leftarrow}(N)$ a disjoint union of homeomorphs of N . A covering space is called a <u>universal covering space</u> if it is simply connected. (If X is a <u>manifold</u>, cf. IV §1.2, then every covering space has a unique manifold structure making the projection differentiable, and there is a unique <u>universal covering manifold</u>, cf. Steenrod [96] pp. 67-71 and IV §2.3 below. Also for manifolds, connectedness implies arcwise connectedness.)

The <u>fundamental group</u> or <u>first homotopy group</u> $\pi_1(X,x_o)$ of a space X with base point $x_o \in X$ is the group consisting of homotopy equivalence classes of closed curves beginning and ending at x_o . If X is arcwise connected then these groups are isomorphic for all choices of x_o . For introductory treatments of the subject see Singer and Thorpe [90] and Wall [100]. We shall not be venturing into algebraic topology but on a couple of occasions we use a covering space for a manifold. Covering spaces can be made from pairs of points and closed curves through the point, by using homotopy equivalence. The interesting result is that, for each subgroup H of a fundamental group $\pi_1(X)$, a covering space $\tilde{X}_H$

can be found with $\pi_1(\tilde{X}_H) = H$.   In particular, if H is trivial
then $\tilde{X}_H$ is simply connected and is therefore the universal
covering space.   Note that Chevalley [12] used the name Poincaré
group for what is now usually called the fundamental group;
physicists use the name Poincaré group for the Lorentz group plus
translations of Minkowski space (cf. DeWitt and DeWitt [23]).   A
general reference work on algebraic topology is Spanier [92].

2.   LIMITING TOPOLOGIES

Given a set X there is a partial order on topologies for X and
we shall use various general properties of this relation.   Detailed
discussions of posets, sets with partial orders, can be found for
example in Birkhoff and Bartree [5] several of whose proofs we
shall adapt for our purposes.

By space we shall mean topological space.

Since we may simultaneously be considering several topologies
on a given set we need to be careful about just calling maps
continuous.   To emphasize which topologies are involved it is
safer to say that a map

$$f : X_1 \longrightarrow X_2$$

from space $(X_1 T_1)$ to (X_2, T_2) is (T_1, T_2) - continuous or to
say that

$$f : (X_1 T_1) \longrightarrow (X_2, T_2)$$

is continuous.

We shall use $\square$ to signify the end of a proof or its
omission.

Our main result is to establish the completeness of the
category Top.   Then we look at adjointness properties.

2.1   *Partial order for topologies*

A partial order for a set S is a reflexive, antisymmetric,
transitive relation $\leqslant$ on S .
This means that :

(i) $\leqslant$ is a subset of S×S, the set of all ordered pairs from S
        (cf. II §2.1 Ex. 3);

(ii)   (x,x) is a member of $\leqslant$ for all x∈S ;

(iii) (x,y) and (y,x) in $\leqslant$ implies x=y;

(iv)   (x,y) and (y,z) in $\leqslant$ implies (x,z) in $\leqslant$.

The word 'partial' is specifically chosen to emphasize that given
a partial order $\leqslant$ on S   and x,y∈S then it is possible that
neither (x,y) nor (y,x) is in $\leqslant$.   When (x,y) is in a partial
order $\leqslant$ we say that x and y are comparable, otherwise they are
incomparable.

       For convenience, we usually abbreviate:

"(x,y) is in $\leqslant$" to "x$\leqslant$y" .

       Given a set X we obtain a partial order $\leqslant$ for topologies
on   X from the partial order $\subseteq$ of subset inclusion by:

$$T_1 \leqslant T_2 \quad <=> \quad (A \in T_1 \implies A \in T_2) .$$

Equivalently

$$T_1 \leqslant T_2 \quad <=> \quad \text{every } T_1\text{-open set in X is } T_2\text{-open also.}$$

We shall write $T_1 < T_2$ if $T_1 \neq T_2$ but $T_1 \leqslant T_2$.

Example 1

Let ρ,τ be two semimetrics for X and suppose for some constant
c > 0, ρ(x,y) $\leqslant$ c τ(x,y), ∀x,y∈X.   Then $T_\rho \leqslant T_\tau$.

Example 2

X_{ind} = {∅,X}, the indiscrete topology and X_{dis} = sub X , the
discrete topology on a set X are the extreme choices, the smallest
and largest respectively.

Remark

Normally when we wish to topologise a set we choose a topology
somewhere between these extremes and our choice is usually
influenced by which maps to or from other spaces are destined to
be continuous.

Example 3

If (Y,T) is any topological space, X is any set and $g : Y \longrightarrow X$, $f : X \longrightarrow Y$ are any maps then we necessarily have:

(i) $g : (Y,T) \longrightarrow (X,X_{ind})$ is continuous,

(ii) $f : (X,X_{dis}) \longrightarrow (Y,T)$ is continuous.

2.2   *Existence of extremal topologies among a finite number*

2.2.1   Theorem

Given any non-empty set $S = \{T_i \mid i = 1,2,\ldots,n\}$ of a finite number n of topologies for a set X there exist minimal and maximal members of S, T_{min} and T_{max}, given by, respectively:

(i) $(\not\exists T_i \in S) : T_i < T_{min}$

(ii) $(\not\exists T_i \in S) : T_{max} < T_i$.

Proof

We use the principle of consistent enumeration for partially ordered sets (cf. e.g. [5] p.40).

Let $S_m = \{T_1, T_2, \ldots, T_m\}$.

We construct inductively a permutation β_m of $1,2,\ldots,m$ such that the re-ordering of S_m as

$$T_{\beta_m(1)}, T_{\beta_m(2)}, \ldots, T_{\beta_m(m)}$$

satisfies

$$T_{\beta_m(i)} < T_{\beta_m(j)} \implies i < j .$$

Necessarily, β_1 is the identity .

Suppose we have β_{r-1} for some $r \geqslant 1$ and let k be the first integer such that

$$T_1 < T_{\beta_{r-1}(i)} = T_{\beta_{r-1}(k)} .$$

Hence we define the permutation

$$\beta_r(i) \;=\; \begin{cases} i & \text{if } i < k \\ r & \text{if } i = k \\ i-1 & \text{if } i > k \end{cases}$$

which puts T_r between $T_{\beta_{r-1}(k-1)}$ and $T_{\beta_{r-1}(k)}$.

So if $\{T_{\beta_r(i)},\, T_{\beta_r(j)}\} \subset S_{r-1}$ then

$$T_{\beta_r(i)} < T_{\beta_r(j)} \implies i < j \; .$$

Also,

$$T_r \;=\; T_{\beta_r(k)} < T_{\beta_r(j)} \implies k < j \; ;$$

and

$$T_{\beta_r(i)} < T_{\beta_r(k)} \;=\; T_r \implies i < k$$

because $T_{\beta_r(k)} < T_{\beta_r(k+1)}$ by construction and so

$$T_{\beta_r(i)} < T_{\beta_r(k+1)} \quad \text{by transitivity of } \leqslant \text{ and then by}$$

construction $i < k + 1$, but since β_r is a bijection we know
that $i \neq k$.

Finally, we apply β_n to the whole set $S = S_n$ and the required
minimal and maximal elements are

$$T_{\min} \;=\; T_{\beta_n(1)}, \quad T_{\max} \;=\; T_{\beta_n(n)} \; . \qquad \square$$

Remark

The generalisation of this that we want for arbitrary sets is like
that needed in the real number system, and the same terminology is
used.

2.3   *Uniqueness of sup and inf topologies*

Given any set S of topologies for a set X we say that :

(i)   a topology T_ℓ for X is a lower bound of S if $T_\ell \leqslant T$
      for all $T \in S$;

(ii)   a topology T_u for X is an <u>upper bound</u> of S if $T \leqslant T_u$
       for all $T \in S$;

(iii) a topology $T_[$ for X is a <u>greatest lower bound</u> of S ,
       written $T_[= \inf S$, if $T_[$ satisfies (i) and also if T_ℓ
       is any other lower bound of S then $T_\ell \leqslant T_[$;

(iv)   a topology $T_]$ for X is a <u>least upper bound</u> of S ,
       written $T_] = \sup S$, if $T_]$ satisfies (ii) and also if
       T_u is any other upper bound of S then $T_] \leqslant T_u$.

Our terminology in (iii) and (iv) implies uniqueness, which we
prove now.

2.3.1   Theorem

For any non-empty set S of topologies for a set X , if they
exist then inf S and sup S are unique.

Proof

We prove the case for inf S, that for sup S is similarly patterned
around the antisymmetry of $\leqslant$.

    Suppose that $T_[$ and $T_($ are both greatest lower bounds for
S . Then $T_[\leqslant T_($ and $T_(\leqslant T_[$. Hence by property §2.1 iii,
$T_[= T_($.   □

    As was to be expected, inf S and sup S do exist; our next
result shows how to construct them.

2.4   *Existence of sup and inf topologies*

2.4.1   Theorem

Given $S = \{T_x | x \in A\}$ an arbitrary set of topologies for a set X ,
there exist unique greatest lower and least upper bounding
topologies

$$T_[= \inf S$$

$$T_] = \sup S .$$

Proof

If $S = \emptyset$ then we have (cf. §2.1 Ex. 1) $T_{[} = X_{ind}$ and $T_{]} = X_{dis}$, unique by §2.3 .

For $S \neq \emptyset$ we need only prove existence because §2.3 guarantees uniqueness. We begin with

$$T_{[} = \{G \in \text{sub } X | \forall \alpha \in A, G \in T_\alpha\} ,$$

the intersection of topologies in the set S , that is by using the open sets common to __all__ members of S . We need to check the axioms §1.1 (i) - (iii). Evidently they hold because each T_α is itself a topology. Hence $T_{[}$ is a topology and plainly a lower bound of S . Suppose that T_ℓ is another lower bound of S . Then we argue:

$$T_\ell \leq T_\alpha \qquad (\forall \alpha \in A) \qquad \text{(by §2.3 i)}$$

$$G \in T_\ell \implies G \in T_\alpha \qquad (\forall \alpha \in A) \qquad \text{(by §2.1)}$$

$$\implies G \in T_{[} , \quad \text{so} \quad T_\ell \leq T_{[} .$$

Next we construct sup S by using as sub base

$$B_{]} = \{G \in \text{sub } X | \alpha \in A : G \in T_\alpha\} ;$$

so $T_{]}$ is the topology generated (cf. §1.2) by arbitrary unions and finite intersections of the members of $B_{]}$. (Of course, the __union__ of topologies in S might not be a topology.) We see that $T_{]}$ is indeed an upper bound of S by :

$$G \in T_\alpha \implies G \in B_{]} \implies G \in T_{]}$$

$$\implies T_\alpha \leq T_{]} \quad (\forall \alpha \in A) .$$

Suppose that T_u is another upper bound. Then we have

$$G \in T_{]} \implies \begin{cases} \text{either } G = B_1 \cap B_2 & \text{for some } B_i \in B \\ \text{or} \quad G = \bigcup_{\lambda \in J} B_\lambda & \text{for some } B_\lambda \in B, \lambda \in J . \end{cases}$$

Hence, either $(\exists \alpha_1, \alpha_2 \in A) : B_i \in T_{\alpha_i}, i = 1,2,$

or $(\exists \alpha_\lambda \in A, \lambda \in J): B_\lambda \in T_{\alpha_\lambda} \forall \lambda \in J .$

In either case we have $G \in T_u$, so $T_u \leq T_{]} .$ □

Remark

We obtain two corollaries that help us see how T_1 is constructed.

Corollary 1

If for each $\alpha \in A$, B_α is a sub base for T_α then $B = \bigcup_{\alpha \in A} B_\alpha$ is a sub base for $T_1 = \sup\{T_\alpha \mid \alpha \in A\}$.

Proof

By the theorem we observe that the members of B are in T_1 . Suppose that $G \in T_1$; we must construct it as a finite intersection or union of sets from B . If $G = \emptyset$ we can intersect the empty set of sets from B . Otherwise we know by construction of T_1 in the theorem that

$$G = \bigcup_{r \in D} G_r \qquad \text{for some indexing set } D$$

with

$$G_r = \bigcap_{i=1}^{r} H_i \qquad \text{for some } H_i \in B_1 \ .$$

Now fix $r \in D$, then for each $i = 1, 2, \ldots, n_r$ there is $\alpha_i \in A$ with $H_i \in T_{\alpha_i}$. So each H_i is a union of finite intersections from the sub base B_{α_i} , hence similarly constructible from B . It follows that G is a union of finite intersections of sets from B . $\square$

Corollary 2

Let $x \in X$ and for each $\alpha \in A$ suppose that $N_\alpha(x)$ is a base for T_α-neighbourhoods of x ; this means that (cf. §1.2) every open set containing x contains a set constructible from finite intersections of sets from $N_\alpha(x)$. Then finite intersections from the set

$$\{N_\alpha(x) \in N_\alpha(x) \mid \alpha \in A\}$$

constitute a base for T_1-neighbourhoods of x .

Proof

The proposed base sets are of the form

$$\bigcap_{i=1}^{n} N_{\alpha_i}(x) \ , \quad \alpha_i \in A, \quad N_{\alpha_i}(x) \in \mathcal{N}_{\alpha_i}(x) \ .$$

Suppose $x \in G_{\alpha_i} \in T_{x_i}$ with $G_{\alpha_i} \subset N_{\alpha_i}(x)$, $i = 1,2,\ldots,n$.

Then $\quad x \in \bigcap_{i=1}^{n} G_{\alpha_i} \subset \bigcap_{i=1}^{n} N_{\alpha_i}(x)$

and

$$\bigcap_{i=1}^{n} G_{\alpha_i} \in T_\rfloor \text{ by construction,}$$

so the proposed sets are indeed $T_\rfloor$ - neighbourhoods of x .

Given some $T_\rfloor$ - neighbourhood $N(x)$ of x we can find from the sub base $B_\rfloor$ some $G_{\alpha_i} \in T_{\alpha_i}$, $i = 1,2,\ldots,n$ for some n and

$$x \in \bigcap_{i=1}^{n} G_{\alpha_i} \subset N(x) \ .$$

But then $x \in G_{\alpha_i}$ for $i = 1,2,\ldots,n$ and so each G_{α_i} is a finite intersection of sets from $\mathcal{N}_{\alpha_i}(x)$. Therefore, $\bigcap_{i=1}^{n} G_{\alpha_i}$ is a finite intersection from

$$\{N_\alpha(x) \in \mathcal{N}_\alpha(x) \mid \alpha \in A\}$$

So we have found a $T_\rfloor$ - neighbourhood of x contained in $N(x)$ and constructed in the form $\bigcap_{i=1}^{n} N_{\alpha_i}(x)$. $\square$

2.5   *Coinduced and induced topologies*

We can construct a unique topology on a set X if we have a map from the set to a topological space or from a topological space to the set. In each case we take the appropriate limiting topology that makes the map continuous.

Lemma

Let (X,T) be a topological space and for some set Y suppose that we have

  either   (i)   a map $f : Y \longrightarrow X$

    or (ii)   a map $g : X \longrightarrow Y$.

Then, in case (i)

$$f^{\leftarrow}T = \{f^{\leftarrow}G \mid G \in T\}$$

is the <u>smallest</u> topology for Y that makes f continuous, and in case (ii)

$$gT = \{H \subseteq Y \mid g^{\leftarrow}H \in T\}$$

is the <u>largest</u> topology for Y that makes g continuous.

<u>Proof</u>

In each case the topology axioms are satisfied because of the way that maps act on intersections and unions. If either Y or X is empty then it has the unique topology $\{\emptyset\}$ and f or g is the empty map. We prove the extremal properties in turn:

(i)   Suppose that T_ℓ is a topology for Y and

   $f : (Y, T_\ell) \longrightarrow (X, T)$ *

   is continuous. By construction, any open set of $(X, f^{\leftarrow}T)$
   is of the form $f^{\leftarrow}G$ for some $G \in T$. Then, by hypothesis *,
   $f^{\leftarrow}G \in T_\ell$ and so

   $$f^{\leftarrow}T \leq T_\ell .$$

(ii) Suppose that T_u is a topology for Y with

   $g : (X, T) \longrightarrow (Y, T_u)$

   continuous. Let $H \in T_u$. Then $g^{\leftarrow}H \in T$ by hypothesis and
   so $H \in gT$ by construction and hence

   $$T_u \leq gT .   \qquad \square$$

<u>Example</u>

The <u>subspace</u> topology on $S \subseteq X$ for a space (X, T) is just the restriction of T to S ,

$$T|_S = \{G \cap S \mid G \in T\} .$$

This coincides with $f^{\leftarrow}T$ where f is the inclusion map of S in X . Observe that f need not be open: a closed interval of R is open in the subspace topology of that interval.

Remark

For the circumstances of the Lemma we call:

(i) $f^{\leftarrow}T$ the coinduced (or inverse image) topology by f from (X,T)

(ii) gT the induced (or quotient) topology by g from (X,T) .

Corollary 1

We collect some properties of the above topologies, for proofs see Császár [20].

(i)   If B is a base (sub base) for T then $f^{\leftarrow}B$ is a base (sub base) for $f^{\leftarrow}T$.

(ii)   If T' is another topology for X with $T' \leqslant T$ then $f^{\leftarrow}T' \leqslant f^{\leftarrow}T$.

(iii) If $\{T_\alpha|\alpha\epsilon A\}$ are topologies for X then $\{f^{\leftarrow}T_\alpha|\alpha\epsilon A\} = f^{\leftarrow} \sup \{T_\alpha|\alpha\epsilon A\}$,

(iv)   A set F is closed in the topology gT

$$<==> g^{\leftarrow}F \text{ is closed in } T .$$

(v)   A map

$$h : (X_1,T_1) \longrightarrow (X_2, T_2)$$

is continuous

$$<==> h^{\leftarrow}T_2 \leqslant T_1$$

$$<==> T_2 \leqslant hT_1 .$$

(vi)   If a map

$$h : (X_1,T_1) \longrightarrow (X_2,T_2)$$

is continuous and open
(or continuous and closed)

$$\text{then } hT_1 = T_2 \qquad \square$$

2.5.1   Theorem

Let $\{(X_\alpha, T_\alpha) \mid \alpha \epsilon A\}$ be a collection of topological spaces and suppose that Y is a set.

(i)   If for each $\alpha \epsilon A$ there is a map $f_\alpha : Y \longrightarrow X_\alpha$, then

$$\sup \{\overleftarrow{f_\alpha} T_\alpha \mid \alpha \epsilon A\}$$

is the _smallest_ topology for which every f_α is continuous.

(ii)   If for each $\alpha \epsilon A$ there is a map $g_\alpha : X_\alpha \longrightarrow Y$, then

$$\inf \{g_\alpha T_\alpha \mid \alpha \epsilon A\}$$

is the _largest_ topology for which every g_α is continuous.

Proof

In each case we know by §2.4 that the candidate exists, is unique and a topology because our Lemma showed that the constituent sets are topologies.   By corollary 1(v) we see that f_α is continuous if and only if

$$\overleftarrow{f_\alpha} T_\alpha \leqslant \sup \{\overleftarrow{f_\alpha} T_\alpha \mid \alpha \epsilon A\}$$

and so we do have the smallest topology in case (i).   Similarly, in case (ii), continuity of g_α is equivalent to

$$\inf \{g_\alpha T_\alpha \mid \alpha \epsilon A\} \leqslant g_\alpha T_\alpha$$

so we have the largest topology here.   □

Corollary 2

(i)   If, for each $\alpha \epsilon A$, B_α is a sub base for T_α then

$$\{\overleftarrow{f_\alpha} G \mid G \epsilon B_\alpha, \ \alpha \epsilon A\}$$

is a sub base for

$$\sup \{\overleftarrow{f_\alpha} T_\alpha \mid \alpha \epsilon A\} \ .$$

(ii)   If for each $\alpha \epsilon A$ there is another topology T_α' for X_α with $T_\alpha \leqslant T_\alpha'$ then $\sup \{\overleftarrow{f_\alpha} T_\alpha \mid \alpha \epsilon A\} \leqslant \sup \{\overleftarrow{f_\alpha} T_\alpha' \mid \alpha \epsilon A\}$ and $\inf \{g_\alpha T_\alpha \mid \alpha \epsilon A\} \leqslant \inf \{g_\alpha T_\alpha' \mid \alpha \epsilon A\} \ .$

Proof

Császár [20] .   □

2.6   *Product and coproduct topologies*

From the definition of product in II §2.2 and its existence in the
category Set, we know that for any collection $\{(X_i, T_i) \mid i \in J\}$ of
topological spaces there exists in Set their product set $\prod\limits_{i \in J} X_i$.
Plainly, the 'right' topology to put on this set is the <u>smallest</u>
one that makes every projection map (existing in Set)

$$\pi_j : \prod_{i \in J} X_i \longrightarrow X_j \quad \text{continuous.}$$

Hence, we define the <u>product</u> topology of the T_i by

$$\prod_{i \in J} T_i \;=\; \sup \{\pi_i^{\leftarrow} T_i \mid i \in J\} \; .$$

2.6.1   <u>Theorem</u>

The <u>product space</u> $\left(\prod\limits_{i \in J} X_i , \; \prod\limits_{i \in J} T_i \right)$ is the product in Top of the

objects $\{(X_i, T_i) \mid i \in J\}$.

Proof

Existence is assured by our previous work.   Also, $\prod\limits_{i \in J} X_i$ is

unique up to isomorphism in Set, being a left limit (cf. II §2.1)
of $\{X_i \mid i \in J\}$,   and $\prod\limits_{i \in J} T_i$ is unique because by §2.3 sup gives a
unique topology.

We need to show that given
any continuous

$$g_i : (K,T) \longrightarrow (X_i, T_i), \; \forall i \in J$$

there is a unique continuous

$$\ell : (K,T) \longrightarrow \left(\prod_{i \in J} X_i , \; \prod_{i \in J} T_i \right)$$

with $g_i = \pi_i \circ \ell \;\; \forall i \in J$.

Our candidate for ℓ is obviously
the one from Set, unique as a map,

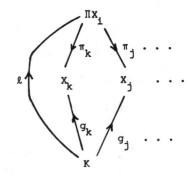

$$\ell : K \longrightarrow \prod_{i \in J} X_i : x \longmapsto (g_i(x))_{i \in J} .$$

It remains to show continuity.

Let G be open in $\prod_{i \in J} X_i$.

Then by §2.4 either

(i) $G = \bigcap_{k=1}^{n} \pi_{i_k}^{\leftarrow} G_{i_k} , \; G_{i_k} \in T_{i_k}$

or

(ii) $G = \bigcup_{i \in J_0} \pi_i^{\leftarrow} G_i , \; G_i \in T_i , \; J_0 \subseteq J .$

In case (i) $\ell^{\leftarrow} G = \bigcap_{k=1}^{n} \ell^{\leftarrow} \circ \pi_{i_k}^{\leftarrow} G_{i_k}$

$= \bigcap_{k=1}^{n} g_{i_k}^{\leftarrow} G_{i_k}$

which is open in K since each g_{i_k} is continuous.

In case (ii) $\ell^{\leftarrow} G = \bigcup_{i \in J_0} \ell^{\leftarrow} \circ \pi_i^{\leftarrow} G_i$

$= \bigcup_{i \in J_0} g_i^{\leftarrow} G_i$

which is also open by continuity of the g_i .   Hence the given

map ℓ is continuous.   □

In a dual manner we define the coproduct topology $\coprod_{i \in J} T_i$ on

a family of spaces $\{(X_i, T_i) | i \in J\}$ to be the largest one that makes
continuous all of the set-theoretic injections (cf. II §2.2)

$$f_j : X_j \longrightarrow \coprod_{i \in J} .$$

That is to say,

$$\coprod_{i \in J} T_i = \inf \{ f_j T_j | j \in J \} .$$

2.6.2   Theorem

The <u>coproduct space</u> or <u>topological disjoint sum</u> $\left(\coprod_{i\in J} X_i, \coprod_{i\in J} T_i\right)$
is the coproduct in Top of the objects $\{(X_i, T_i) \mid i\in J\}$.

Proof

Existence and uniqueness in Top follow as for the previous theorem.
We must show that our candidate has the right limit universal
property (cf. II §2.1) : given any continuous maps

$$g_i : (X_i, T_i) \longrightarrow (K,T), \; \forall i\in J ,$$

there exists a unique continuous map

$$r : \left(\coprod_{i\in J} X_i, \coprod_{i\in J} T_i\right) \longrightarrow (K,T)$$

such that

$$g_i = r\circ f_i \quad \forall i\in J .$$

From Ex. 2.1.4 we already have a
candidate for r, it is the unique   map

$$r : f_i(x) \longmapsto g_i(x) \quad \text{if} \quad x\in X_i \overset{f_i}{\hookrightarrow} \coprod_{i\in J} X_i .$$

We need only show continuity.

Suppose then that G is open in K; we must show that $r^{\leftarrow}G$ is
open in $\coprod_{i\in J} X_i$.

Since each f_i is continuous by construction, we have:

$$f_i^{\leftarrow}G = G_i\in T_i \quad \forall i\in J ,$$

and by construction of r

$$f_i^{\leftarrow}\circ r^{\leftarrow}G = g_i^{\leftarrow}G .$$

So, as required, we have shown that, for each $i\in J$, $f_i^{\leftarrow}(r^{\leftarrow}G)$ is
open in X_i because each g_i is continuous by hypothesis.   □

Remark

It is of course important to know which properties of constituent
spaces persist in the product and coproduct spaces.   We collect a
few of these properties, for proofs and more results see e.g.

Császár [20] and for counterexamples Steen and Seebach [95].

Property of all factor spaces (X_i, T_i) i∈J	Whether preserved by product and coproduct					
	J finite		J countable		J uncountable	
	Π	⊥⊥	Π	⊥⊥	Π	⊥⊥
Compact	Yes	Yes	Yes	No	Yes	No
Locally compact	Yes	Yes	No	Yes	No	Yes
First countable	Yes	Yes	Yes	Yes	No	Yes
Second countable	Yes	Yes	Yes	Yes	No	No
Separable	Yes	Yes	Yes	Yes	No	No
Paracompact	No	Yes	No	Yes	No	Yes
T_0, T_1, T_2 = Hausdorff	Yes	Yes	Yes	Yes	Yes	Yes
Connected	Yes	No	Yes	No	Yes	No

2.7   *Completeness of Top*

We know from Ch. II §2.5 Ex. 9 (cf. Higgins [47] p.54) that a category is left complete if it admits products and equalizers and right complete if it admits coproducts and coequalizers. We have established products and coproducts in Top, in the previous section.   Also, in Ch. II §2.4 Exs. 1,2 we constructed equalizers and coequalizers in Set, next we provide them with their natural topologies.

2.7.1   Theorem

Top is a complete category.

Proof

After §2.6 we need only establish equalizers and coequalizers. So consider in Top the diagram

$$\Delta : (X_1, T_1) \quad \underset{a_2}{\overset{a_1}{\longrightarrow}} \quad (X_2, T_2)$$

(i)   The equalizer set is
      (cf. II §2.4)

      $X = \{x \in X_1 \mid a_1(x) = a_2(x)\}$

      and we have to provide it
      with a topology that makes
      continuous the inclusion

        $f_1 : X \hookrightarrow X_1$,

which is an injective set map.

We know that a_1 is continuous so then $f_2 = a_1 \circ f_1$ will be
continuous.   Hence we need the subspace topology, coinduced
by f_1 (cf. §2.5) so our candidate for $\underleftarrow{\lim} \Delta$ in Top is the
space $(X, f_1^{\leftarrow}T)$.   Suppose that we have continuous maps

$$(K, T_o) \overset{g_1}{\longrightarrow} (X_1, T_1)$$
$$\overset{g_2}{\searrow} (X_2, T_2)$$

commuting with the a_1, a_2.   We have the map:

     $\ell : K \longrightarrow X : y \longmapsto y'$

where $\{y'\} = f_1^{\leftarrow} g_1(y)$

because by commutativity a_1 and a_2 agree on X in the
image of g_1.   We need to establish continuity of ℓ.   Let
G be open in $(X, f_1^{\leftarrow}T_1)$.   Then we have some $S \in T_1$ with
$G = f_1^{\leftarrow}S$ by Lemma 2.5.   Thus we deduce

     $\ell^{\leftarrow}G = \ell^{\leftarrow} \circ f_1^{\leftarrow}S = g_1^{\leftarrow}S$ (since $f_1 \circ \ell = g_1$)

which is open in K since g_1 is continuous.

(ii) The coequalizer set is (cf. II §2.4)

     $Y = X_2/\sim$

where $\sim$ is the smallest equivalence relation on X_2 that
contains

$$\{(a_1(x), a_2(x)) \mid x \epsilon X_1\} .$$

Also,

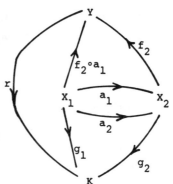

$$f_2 : X_2 \longrightarrow Y : x \longmapsto [x]_\sim$$

is the surjection onto classes
and

$$f_1 = f_2 \circ a_1 .$$

The right topology is that induced
by f_2 (cf. §2.5) so our candidate
right limit in Top is $(Y, f_2 T_2)$.

Suppose that we are given
continuous maps

$$
\begin{array}{ccc}
(X_1, T_1) & & (X_2, T_2) \\
& \searrow g_1 \quad g_2 \swarrow & \\
& (K, T) &
\end{array}
$$

commuting with the a_1, a_2.   Then we already have in Set the unique
map

$$r : Y \longrightarrow K : [x]_\sim \longmapsto g_2(x)$$

satisfying $r \circ f_2 = g_2$, $r \circ f_1 = r \circ (f_2 \circ a_1) = g_1$.
This map is well-defined for suppose that

$$y \epsilon [x]_\sim$$

then

$$a_1(x) = a_2(y) \quad \text{and so} \quad g_2 \circ a_1(x) = g_2 \circ a_2(y)$$

or $\qquad a_1(x) = a_1(y) \qquad \text{so} \qquad g_2 \circ a_1(x) = g_2 \circ a_1(y)$

or $\qquad a_2(x) = a_2(y) \qquad \text{so} \qquad g_2 \circ a_2(x) = g_2 \circ a_2(y)$

hence $r([x]_\sim) = r([y]_\sim)$ whenever $(x \sim y)$.

It remains to show that r is continuous.   Let G be open in
(K,T) .   Now, by §2.5, H is open in $(Y, \overset{\leftarrow}{f_2} T_2)$ only if $\overset{\leftarrow}{f_2} H$

is open in T_2 .   But

$$f_2^{\leftarrow}(r^{\leftarrow}G) = (r \circ f_2)^{\leftarrow}G = g_2^{\leftarrow}G$$

and so $r^{\leftarrow}G$ is indeed open in $(Y, f_2^{\leftarrow}T_2)$.

Therefore, Top admits coequalizers and by (i) equalizers; since by §2.6 it also admits products and coproducts Top is complete : it has left and right limits for all of its diagrams. ▢

Remark

We have previously noted that if a left (or right) limit exists in a category C then it is unique up to isomorphism in C .   In Top, an isomorphism is a homeomorphism, a bijective continuous map with continuous inverse.

2.8   *Projective limit and inductive limit topologies*

Products and coproducts arise from collections of isolated spaces. In general we are interested in collections of spaces with some maps among them;   then we can use products and coproducts as the basis for constructing limits.

Our concern here is with limits of diagrams in Top like

$$\Delta = \{p_{ij} : (X_i, T_i) \longrightarrow (X_j, T_j) \mid i \leq j,\ i,\ j \in J\}$$

where (i)   J is some non-empty set of indices with partial
             order $\leq$

(ii) $p_{ii} = I_{X_i}$ $(\forall i \in J)$

(iii) $p_{ik} = p_{jk} \circ p_{ij}$ whenever $i \leq j \leq k$.

The projective limit space is the left limit of Δ and the inductive limit space is the right limit of Δ .   These spaces exist by the completeness of Top (cf. §2.7) and, like all limits there, are unique up to homeomorphism.   Further discussion will be found in Kowalsky [59].   As we pointed out in Ch. II §2.5 Ex. 8, set products and coproducts lie at the basis of limits in Top and we shall prove this next for the present cases.   We shall see in Chapter V how the projective limit is used to construct a boundary for spacetime (cf. [15], [26]).

2.8.1   Theorem

With Δ as above, the left limit object is given by the subspace
(cf. §2.6) of the product space

$$X = \{(x_i)_{i\in J} \in \prod_{i\in J} X_i \,|\, j{\leq}k{\in}J \Longrightarrow p_{jk}(x_j) = x_k\}$$

with associated morphisms the restrictions of the projections
(cf. §2.6 and II §2.2)

$$p_k = \pi_k\big|_X \quad : \quad (x_i)_{i\in J} \longrightarrow x_k \; .$$

Proof

Existence is assured by completeness of Top (cf. §2.7) and so we
only need to establish the
left limit property.   Suppose
then that we have continuous
maps

$$g_i : (K,T) \longrightarrow (X_i,T_i)$$

for all $i{\in}J$, commuting with
the contiguous p_{jk} .

From §2.6 we have a continuous
map

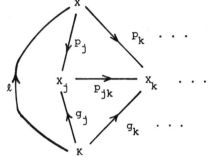

$$\ell : K \longrightarrow \prod_{i\in J} X_i : x \longrightarrow (g_i(x))_{i\in J} \; .$$

We can see that for the given g_i it takes values in X.   For if
$j{\leq}k$ then we deduce

$$(\forall x{\in}K) \; p_{jk}{\circ}g_j(x) = g_k(x) \; ,$$

by commutativity. $\Box$

2.8.2   Theorem

With Δ as above, the right limit object is given by the set of
equivalence classes

$$Y = \{[x]_\sim \,|\, x \in \coprod_{i\in J} X_i \}$$

where $\sim$ is the relation

$$x \sim y \iff \exists i \leq k, \ j \leq k \quad \text{with} \quad p_{ik}(x) = p_{jk}(y)$$

and associated morphisms are the projections onto the classes (cf. II §2.2) :

$$f_i : X_i \longrightarrow Y : x \longmapsto [x]_\sim$$

and the topology is that induced by projection onto classes from the space

$$\left(\coprod_{i \in J} X_i \ , \ \coprod_{i \in J} T_i \right) \ .$$

Proof

We know by §2.7 that a right limit exists. We show that our candidate meets the requirements, then we know that it is unique up to homeomorphism.

Suppose that we have continuous maps

$$g_i : (X_i, T_i) \longrightarrow (K, T)$$

for all $i \in J$, commuting with the contiguous p_{jk}. From §2.6 we have a continuous map

$$r : \coprod_{i \in J} X_i \longrightarrow K : f_j(x) \longmapsto g_j(x)$$
$$\text{when } x \in X_j \ .$$

Next we note that $\sim$ is an equivalence relation.   The proposed topology on Y is induced by

$$q : \coprod_{i \in J} X_i \longrightarrow Y : x \longmapsto [x]_\sim \ ,$$

which is to say that

$$q : \left(\coprod_{i \in J} X_i, \ \coprod_{i \in J} T_i \right) \longrightarrow \left(Y, \ q \coprod_{i \in J} T_i \right)$$

is continuous.   It follows that the composite

$$q \circ r : \left(Y, \ q \coprod_{i \in J} T_i \right) \longrightarrow (K, T)$$

is also continuous and satisfies

$$g_j = q \circ r \circ f_j \quad \forall j \in J \ .$$

If also there is a continuous map

$$s : \left(Y, \; q \coprod_{i \in J} T_i\right) \longrightarrow (K,T)$$

satisfying

$$g_j \;=\; s \circ f_j \quad \forall j \in J \; ,$$

then

$$(\forall y \in Y)\,(\exists j \in J, \; x \in X_j) : \; f_j(x) \;=\; [x]_{\sim} \;=\; y \; .$$

Hence

$$g_j(x) \;=\; s[x]_{\sim} \;=\; s(y)$$

but then

$$q \circ r(y) \;=\; q \circ r(f_j(x) \;=\; g_j(x) \;=\; s(y) \; .$$

So the map $q \circ r$ is unique.   □

Example 1

(Cf. Császár [20] p.297, Ex. 2)

Let (X_o, T_o) be a topological space, let $J = N = \{1,2,\ldots\}$ with
the usual ordering, and suppose we have subsets $\{X_i \,|\, i \in N\}$ of X_o
giving the inclusion diagram

$$\emptyset \neq X_1 \;\underset{p_{12}}{\overset{\hookrightarrow}{}}\; X_2 \;\underset{p_{23}}{\overset{\hookrightarrow}{}}\; X_3 \;\cdots\; X_i \;\underset{p_{ii+1}}{\overset{\hookrightarrow}{}}\; X_{i+1} \;\cdots$$

for subspaces $\{(X_i, T_i) \,|\, i \in N\}$, so $T_i = T_o|_{X_i}$.

Then the projective limit space is homeomorphic to the subspace
$\bigcap\limits_{i=1}^{\infty} X_i$ of X_o

Example 2

(Cf. Császár [20] p. 330, Ex. 18)

Let (X_o, T_o) be a topological space and $\{X_i \,|\, i \in J\}$ any collection
of subsets of X_o , each X_i having the subspace topology T_i .
Then (X_o, T_o) is homeomorphic to the inductive limit of the
inclusion diagram.

IV Manifolds and bundles

We can think of a differentiable structure for a topological space as the provision for analysis on that space. Quite generally, a derivative of a map f is a linear map (a morphism from Vec) that locally is an approximation to f itself. This is a natural abstraction from real analysis and the least extra structure needed to support it is a consistent establishment of local tangent vector spaces. A differentiable manifold is a topological space on which such derivatives can be defined. An introductory treatment of manifolds can be found in Dodson and Poston [28], which begins with vector spaces and proceeds to a rigorous formulation of relativistic spacetime; mainly our notation will follow this text. Manifolds and differentiable maps form a category, Man, but it is not complete because of the absence of certain pullbacks, quotients and coproducts (cf. Lang [60] and Hirsch [49]).

Some superstructure is inevitably present on a differentiable manifold, namely its tangent bundle, which is a differentiable manifold made up from the local tangent vectors. In applications, other such vector bundles are useful, and indeed so are the more general fibre bundles which do not break down into local vector spaces but into local differentiable manifolds. Thus, all scalar, vector and tensor fields in physics are sections (cf. II §1.10, Ex. 4) of appropriate vector bundles over a spacetime manifold. On the other hand, a parallelization (a smooth choice of basis for tangent vector spaces over a manifold, cf. §2.4 below) is a section of the fibre bundle of linear frames (ordered bases for tangent spaces).

In this chapter we shall collect some basic terminology for
manifolds and bundles and provide for certain constructions that
we shall need for spacetime geometry in the next chapter.

There are many excellent texts on manifolds and bundles so we
shall not attempt more than an introductory guide to the subject.
Basic manifold theory will be found in an eminently digestible
form in Spivak [94], Brickell and Clark [9], Bishop and Crittenden
[6], and more advanced material is given in Kobayashi and Nomizu
[57, 58], Lang [60], Schwartz [87] and Hirsch [49].   Several of
these books make some use of bundles but for a detailed treatment
of bundle theory see Husemoller [52] and of course the classic
Steenrod [96].   Intimately connected with the bundle geometry
constructed over spacetime manifolds is the notion of a Lie group,
being a group that is also a manifold.   A detailed account of the
theory of these entities is given by Hochschild [50].
Applications of bundle theory to the analysis of spacetime
singularities are studied in Dodson [26], to which we shall refer
in the sequel.

1.   MANIFOLD STRUCTURE

1.1   *Topological vector space, differentiation, tensor spaces;*
   exact sequences

Given a real vector space V of dimension n we immediately have
the dual vector space $V^* = L(V;\mathbb{R})$ of all real valued linear maps
on V . The natural choice of topology for V is (cf. III §2.5)

$$T_V = \sup \{f^{\leftarrow} T_{\mathbb{R}} | f \in V^*\}$$

where $T_{\mathbb{R}}$ is the standard metric topology for $\mathbb{R}$.   It turns out
that this topology for V has other equivalent formulations
(cf. [28]).   For example, we could take any basis
$\{b_i | i = 1,2,\ldots,n\}$ for V and obtain a family of maps
$\{b^i | i = 1,2,\ldots,n\}$

making the diagram commute.   Here b is an isomorphism in Vec
and the π^i are the projections from the product in Top.
Evidently the b^i form a basis dual to the b_i.   Since finite
linear combinations of real valued maps are continuous we could
topologise V by using

$$T'_V = \sup\{b^{i^+}T_R \mid i = 1,2,\ldots,n\} \ .$$

However, $T'_V = T_V$ (cf. [28] for details and pictures of open sets).
We call T_V the <u>usual topology</u> for V .   Similarly, we shall
always understand that vector spaces of linear maps, like L(V;W),
have the usual topology.   This all fits together with the familiar
situation of the linear maps from R^n to R^m being represented
by matrices in $R^{m\times n} = R^{mn}$.

   The <u>tangent space</u> at $x\epsilon V$ is the vector space

$$T_x V = \{x\}\times V$$

which is isomorphic to V (in Vec and Top).   Given any map (not
necessarily linear)

$$f : V \longrightarrow W$$

between vector spaces V and W then the <u>derivative</u> of f at
$x\epsilon V$ (if it exists it is unique, cf. [28]) is a <u>linear map</u>

$$D_x f : T_x V \longrightarrow T_{f(x)}W$$

such that:

   for any neighbourhood of $\underline{0} \epsilon L(T_x V; T_{f(x)}W)$

   there is a neighbourhood N' of $\underline{0} \epsilon T_x V$

for which

$$(\forall(x,t)\epsilon N') (\exists A\epsilon N) : (f(x),f(x+t)) = D_x f(x,t) + A(x,t) \ .$$

Now, if dim V = n,   dim W = m   then, once we choose bases

for $T_x V \equiv V$ and $T_{f(x)} W \equiv W$, the linear map $D_x f$ is represented

by a unique matrix $[\partial_i f^j]_x$ making the following diagram commute

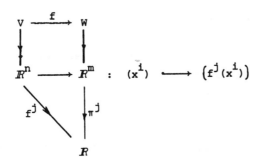

Of course, the entries in $[\partial_i f^j]_x$ are the partial derivatives

the constituent maps of f via the chosen bases :

A map f : V $\longrightarrow$ W is <u>differentiable at x</u> if $D_x f$ exists,

and f is called <u>differentiable</u> if $D_x f$ exists for all $x \in V$.

Since $T_x V$ and $T_{f(x)} W$ have their usual topologies we know

that the linear map $D_x f$ is continuous.   For handling higher

derivatives it is useful to introduce (for differentiable f) the

map

$$\hat{D}f \ : \ V \longrightarrow L(V;W)$$

$$: \ x \longmapsto \hat{D}_x f$$

where

Then we say : f is C^1 if $\hat{D}f$ is continuous; f is C^2 if $\hat{D}f$
is C^1 , etc. If f is C^k for all finite k then we say
that f is C^∞ or <u>smooth</u> .

For a detailed discussion of the above and for the more general
definition of differentiation on affine spaces see [28] Ch. 7, and
Porteous [78] Ch. 18.   In [28] an appendix provides a modern proof
of the inverse function theorem through the proof of a theorem on
the existence and uniqueness of smooth flows (on a manifold).   The
classical form for the inverse function theorem for $f : R^n \longrightarrow R^n$
is:

1.1.1   <u>Theorem</u>

If f is C^k for any k then $D_x f$ is an isomorphism if and only
if there are neighbourhoods N of x, N' of f(x) with f(N) = N'
and there is a local C^k inverse $\overleftarrow{f}$: N' $\longrightarrow$ N . $\square$

<u>Corollary</u>

If $f : R^n \longrightarrow R^m$ (n≤m) is C^1 and $D_x f$ is injective then
there is a neighbourhood N of x such that $f|_N$ is injective. $\square$

In these assertions the domain of f can also be any non-
empty set in R^n ; both carry over to maps between manifolds.

For physics it turns out that real valued linear functions of
several vector variables are important, the <u>multilinear functionals</u>
or <u>tensors</u>. A detailed motivation for the following definition
will be found in [28] Ch. V.   A <u>tensor product</u> of vector spaces
X_1 , ..., X_n is a vector space X together with a multilinear map

$$\otimes : X_1 \times X_2 \times \ldots \times X_n \longrightarrow X$$

such that if also there is a multilinear map

$$f : X_1 \times X_2 \times \ldots \times X_n \longrightarrow Y$$

then there is a unique linear map

$$\hat{f} : X \longrightarrow Y \text{ with } f = \hat{f} \circ \otimes .$$

Diagrammatically (cf. II §2.2) we require commutativity of

It follows that the tensor product always exists and is unique up
to isomorphism (cf. [28] p. 146).   The actual construction of

$$X = X_1 \otimes X_2 \otimes \ldots \otimes X_n$$

is quite simple.   First we define (cf. II §1.4 for the introduction
of dual vector spaces):

$$X_1^* \otimes X_2^* \otimes \ldots \otimes X_n^* = L(X_1, X_2, \ldots, X_n; R) ;$$

then we exploit the finite-dimensional natural isomorphism of
taking double duals (cf. II §1.7) to put $X_i^{**} \equiv X_i$ and hence choose

$$X_1 \otimes X_2 \otimes \ldots \otimes X_n = L(X_1^*, X_2^*, \ldots, X_n^*; R) .$$

Example

Given any vector spaces X_1, X_2 and their duals X_1^*, X_2^* , typical
elements are

$$x \in X_1, \ y \in X_2, \ f \in X_1^*, \ g \in X_2^*$$

and then

$$x \otimes y : X_1^* \times X_2^* \longrightarrow R : (f,g) \longmapsto f(x)\, g(y)$$

is an element of $X_1 \otimes X_2$ (and every element of $X_1 \otimes X_2$ is a
linear combination of such individual products, though the map $\otimes$
so defined is not surjective, cf. [28]).   Similarly,

$$f \otimes g : X_1 \times X_2 \longrightarrow R : (x,y) \longmapsto f(x)\, g(y)$$

is an element of $X_1^* \otimes X_2^*$.

     Formally, the tensor product is a __bifunctor__ or functor of two
variables (cf. II §2.7):

where

$$f \otimes g \left(\sum_i a_i \otimes b_i \right) = \sum_i \left(f(a_i) \otimes g(b_i) \right) .$$

Of course this construction can equally well be performed in the category of Abelian groups. In general, the adjoint functor theorem (cf. Ch. II §2.8) provides for the construction of tensor products. Further discussion will be found in MacLane [67] p.159-160 and p.222.

Certain diagrams in the category Vec have particular significance, these are the <u>exact sequences</u>. The notion extends to other categories, notably to subcategories of Ab (cf. Herrlich and Strecker [46] §39 and MacLane [67] p.196) and, as we shall see below, to vector bundles (cf. Lang [60] p.48).

A diagram in Vec of the form

$$\cdots V_{-1} \xrightarrow{f_{-1}} V_0 \xrightarrow{f_0} V_1 \xrightarrow{f_1} V_2 \cdots$$

(whether finite or infinite) is called a <u>sequence</u> and it is called an <u>exact sequence</u> if

$$(\forall k) \quad \text{im } f_k = \text{ker } f_{k+1} ;$$

this means

$$(\forall k) \quad f_k(V_k) = \overset{\leftarrow}{f}_{k+1}\{\underline{0}\} .$$

For any vector space V we usually omit names for the unique maps involving a trivial space:

$$\{\underline{0}\} \longrightarrow V \quad : \quad \underline{0} \longmapsto \underline{0}$$

$$V \longrightarrow \{\underline{0}\} \quad : \quad x \longmapsto \underline{0}$$

Examples

(Cf. Porteous [78] p.90-95 for basic theory.)

1. $\{\underline{0}\} \longrightarrow W \xrightarrow{s} X$ is exact $\Longleftrightarrow$ s injective .

2. $X \xrightarrow{t} Y \longrightarrow \{\underline{0}\}$ is exact $\Longleftrightarrow$ t surjective .

3. $\{\underline{0}\} \longrightarrow X \xrightarrow{t} Y \longrightarrow \{\underline{0}\}$ is exact

$\Longleftrightarrow$ t is an isomorphism

4. $\{\underline{0}\} \longrightarrow X \xrightarrow{i} X \times Y \xrightarrow{q} Y \longrightarrow \{\underline{0}\}$

and $\{\underline{0}\} \longrightarrow Y \xrightarrow{j} X \times Y \xrightarrow{p} X \longrightarrow \{\underline{0}\}$

(with i and j the obvious injections and q,p the obvious projections) are exact with $p \circ i = I_X$ and $q \circ j = I_Y$.

5. Given a vector subspace W of a vector space X ,

$$\{\underline{0}\} \longrightarrow W \hookrightarrow X \xrightarrow{\pi} X/W \longrightarrow \{\underline{0}\}$$

is exact. Recall that the factor space, or quotient space is defined by:

$$X/W = \{[x]_W | x \in X\} ,$$

$$[x]_W = \{x + w | w \in W\} = \pi(x) \ (\forall x \in X) ,$$

$$\lambda[x]_W = [\lambda x]_W , \quad \forall \lambda \in R,$$

$$[x]_W + [y]_W = [x+y]_W .$$

An exact sequence of the form

$$\{\underline{0}\} \longrightarrow W \xrightarrow{s} X \xrightarrow{t} Y \longrightarrow \{\underline{0}\}$$

is called a short exact sequence and in view of Example 5 it is often helpful to think of W as a subspace of X , and of Y as X/W . Such a short exact sequence is said to be a split exact sequence if there is given a linear section of t (cf. II §1.10) :

$$t' : Y \longrightarrow X \quad \text{with} \quad t \circ t' = I_Y .$$

1.2   *Manifold definitions, universal cover*

A C^k manifold of dimension n is a Hausdorff topological space
M with a collection of open sets

$$\{U_\alpha \mid \alpha \in A\}$$

in M and corresponding maps

$$\phi_\alpha : U_\alpha \longrightarrow \mathbb{R}^n$$

such that:

(i) $\underset{\alpha \in A}{\cup} U_\alpha = M$.

(ii)   Each ϕ_α defines a homeomorphism $U_\alpha \longrightarrow \phi_\alpha(U_\alpha)$.

(iii) If $U_\alpha \cap U_\beta \neq \emptyset$ then the composites

$$\phi_\alpha \circ \overset{\leftarrow}{\phi_\beta} : \phi_\beta(U_\alpha) \longrightarrow \phi_\alpha(U_\beta)$$

$$\phi_\beta \circ \overset{\leftarrow}{\phi_\alpha} : \phi_\alpha(U_\beta) \longrightarrow \phi_\beta(U_\alpha)$$

are C^k maps (between subsets of $\mathbb{R}^n$) .

We call each pair (U_α, ϕ_α) a chart and the set
$\{(U_\alpha, \phi_\alpha) \mid \alpha \in A\}$ an atlas for M .   The properties of M are not
significantly altered if we add to such an atlas more charts
satisfying the requirements (i) - (iii) .

Unless we specify to the contrary we shall mean by 'manifold'
a smooth (i.e. C^∞) manifold.

1.2.1   Theorem

Given a connected manifold M there is a unique universal covering
manifold $\tilde{M}$; that is, a unique universal covering space (cf. III
§1.7) with manifold structure.

Proof

Steenrod [96] pp. 67-71.   □

Example

Real projective n-space $\mathbb{R}\text{P}^n$ is the quotient of the sphere

$$S^n = \{x \in \mathbb{R}^{n+1} \mid \|x\| = 1\}$$

by the group

$$G = \{I, J\} \quad \text{where} \quad \begin{cases} I : x \longmapsto x \\ \\ J : x \longmapsto -x . \end{cases}$$

It turns out that $S^n = \widetilde{\mathbb{R}P}^n$, the sphere is the universal covering manifold of projective space.

1.3   Tangent space and derivatives : manifold category

A map $f : M \longrightarrow N$ between manifolds is called differentiable (or C^r) at $x \in M$ if for some charts (U, ϕ) on M , (V, ψ) on N with $x \in U$, $f(x) \in V$ the map

$$\psi \circ f \big|_U \circ \phi^{\leftarrow} : \phi(U) \longrightarrow \psi(V)$$

is differentiable (or C^r) at $\phi(x)$.
If f is differentiable (or C^r) at all $x \in M$ then we just call f differentiable (or C^r) .

It turns out that this property is independent of the choice of charts at x and $f(x)$.   Smooth manifolds and smooth maps between them form a category Man;   isomorphisms in this category are called diffeomorphisms.

The tangent space to M at x is denoted $T_x M$ and constructed as an isomorph in Vec of the tangent space $T_{\phi(x)} \mathbb{R}^n$; two equivalent constructions are discussed in detail in [28].   So if f as above is differentiable at x then we obtain a linear map, the derivative of f at x :

$$D_x f : T_x M \longrightarrow T_{f(x)} N .$$

Further, by means of the local co-ordinates induced by the chart maps

$$\phi : U \longrightarrow \mathbb{R}^n : x \longmapsto (x^1, x^2, \ldots, x^n)$$

$$\psi : V \longrightarrow \mathbb{R}^m : y \longmapsto (y^1, y^2, \ldots, y^m)$$

the map $D_x f$ is uniquely determined by the Jacobian matrix $[\partial_i f^j]_x$, which is the derivative at $\phi(x)$ of

$$\psi \circ f|_U \circ \overset{\leftarrow}{\phi} : R^n \longrightarrow R^m : (x^i) \longmapsto \left(f^j(x^1, x^2, \ldots, x^n) \right)$$

The <u>rank</u> of f at x is defined to be the rank of the linear map $D_x f$; it coincides with the matrix rank of $[\partial_i f^j]_x$.

If the rank of f at $x \in M$ is maximal then we say that x is a <u>regular</u> point of f , otherwise x is a <u>singular</u> point of f . The important result is due to Sard:

1.3.1   Theorem

If f : M $\longrightarrow$ N is smooth and S is the set of singular points of f then fS has measure zero.

Proof

Schwartz [87] p.9.   Note that S is said to have measure zero if for all charts (U, ϕ)

$$\phi(U \cap S) \text{ has Lebesgue measure zero in } R^n . \qquad \Box$$

(Also in Schwartz [87] p.7, by the way, will be found proof that every connected one-dimensional manifold is diffeomorphic to either an open interval or a circle.   These two results yield the Brouwer Fixed-Point Theorem.)

A <u>curve</u> in a manifold M is a map (C^1 unless otherwise stated)

$$c : I \longrightarrow M : t \longmapsto c(t)$$

from some interval $I \subseteq R$.   The <u>tangent vector to</u> c at $t \in I$ is denoted $\dot{c}(t)$ and defined by $\dot{c}(t) = D_t c(1)$, where 1 is the standard unit vector in $T_t R$.

1.4   Submanifold, product and quotient manifolds

A subset S of a manifold M is a <u>submanifold</u> of M if the inclusion map i : S $\hookrightarrow$ M is an <u>immersion</u>, that is if the rank of i equals the dimension of S .   This of course depends on providing first a manifold structure for S , the construction can be found for example in Lang [60] p.24.

The <u>product manifold</u> of two manifolds M,N is the product topological space M×N (cf. III §2.6) provided with the obvious

differentiable structure from the products of chart maps.
Evidently this process easily extends to any finite number of
factors; we shall not need infinite product manifolds. The
dimension of a product manifold is the sum of the dimensions of
its constituent factors.

A differentiable map $f : M \longrightarrow N$ between manifolds is
called a <u>submersion</u> if its rank equals the dimension of N at
each point of its domain.
Then $(\forall x \in M)$ $D_x f$ is surjective.

Given an equivalence relation ρ on a manifold M we can
take the set of equivalence classes

$$X = M/\rho = \{[x]_\rho \mid x \in M\}$$

and so obtain a surjection

$$g : M \longrightarrow M/\rho : x \longmapsto [x]_\rho .$$

If M/ρ is provided with a differentaible structure that makes g
a submersion then we say that M/ρ is a <u>quotient manifold</u> of M .
Often an equivalence relation arises on M from the action of a
group, as we shall see below. Brickell and Clark [9] Ch. 6
discusses quotient manifolds in some detail and provides examples.

A number of manifolds arise as subsets of R^n for some n .
The following theorem gives a large family of these.

1.4.1   <u>Theorem</u>

Let $f : R^n \longrightarrow R$ be smooth and let $M = f^{\leftarrow}\{0\}$. If at each
point $x \in M$ f has rank one then M admits a C^∞ manifold
structure of dimension n-1.

<u>Proof</u>

The construction of the atlas depends on the implicit function
theorem and full details can be found in Brickell and Clark [9]
p.25.   □

The manifold so constructed is called a <u>differentiable variety</u>
and the spheres are examples. As another example we have the set
of all non-singular n×n matrices with real elements; this is
evidently the differentiable variety (in R^{n^2}) given by the
determinant function. It is important because it is also a group

with smooth operations.

1.5   *Lie group : action : transitive, free, effective*

A **Lie group** is a group G that is also a smooth manifold such
that the group operation

$$G \times G \longrightarrow G : (a,b) \longmapsto ab^{-1}$$

is smooth.   This means that the following maps are smooth for all
$a \in G$

$$L_a : G \longrightarrow G : b \longmapsto ab \quad \text{(left translation)}$$

$$R_a : G \longrightarrow G : b \longmapsto ba \quad \text{(right translation)}$$

$$^{-1} : G \longrightarrow G : b \longmapsto b^{-1} \quad \text{(inversion)} .$$

We shall use e to denote the identity element in G.

Example 1

The **general linear group** $G\ell(n;R)$ of all n×n real non-singular
matrices is an open submanifold of R^{n^2}.   We shall write

$$R^* = G\ell(1;R) .$$

Example 2

The group of real matrices:

$$\left\{ \begin{bmatrix} \beta & v\beta \\ \\ v\beta & \beta \end{bmatrix} \text{ with } 1 > v^2 \text{ and } \beta = (1-v^2)^{-\frac{1}{2}} \right\}$$

is an important Lie group in relativity, as a one-parameter
subgroup of the **Lorentz group** (cf. Porteous [78] p.161).   It has
an equivalent representation as the matrix group: (cf. Gel'fand,
Minlos and Shapiro [35] p.184 et seq.):

$$\left\{ \begin{bmatrix} \cosh\chi & \sinh\chi \\ \\ \sinh\chi & \cosh\chi \end{bmatrix} \text{ with } \chi \in R \right\} .$$

Let G be a Lie group and P a manifold.   Then G **acts**
on P **to (or on) the right** if there is a map

$$P \times G \longrightarrow P : (u,g) \longmapsto R_g(u)$$

satisfying

(i) $(\forall g \in G)$ $g : P \longrightarrow P : u \longmapsto R_g(u)$ is a diffeomorphism

(ii) $(\forall g, h \in G, \forall u \in P)$ $R_{gh}(u) = R_h \circ R_g(u)$.

Example 1

Take P = G and use right translation in G :

$$G \times G \longrightarrow G : (h,g) \longmapsto R_g(h) = hg .$$

An action of a Lie group G on the right of a manifold P
is called

(i)   <u>transitive</u> if $(\forall u, v \in P)$ $\exists g \in G : R_g(u) = v$;

(ii) <u>free</u> if the only element of G with a fixed point is e ;

(iii) <u>effective</u> if

$$(R_g(u) = u \; \forall u \in P) \implies g = e .$$

Example 2

Take $G = R^*$, the multiplicative group on $R \backslash \{0\}$ and $P = R^2$
then

$$R^2 \times R^* \longrightarrow R^2 : \big((x,y),g\big) \longmapsto (gx,gy)$$

is not transitive.   However, right translation in any G (cf.
Example 1) is always transitive because

$$x,y \in G \implies (\exists g = x^{-1}y) : R_g(x) = y .$$

Example 3

Let $P = S^1 \times R^*$.   Then the action induced on P by right
translation in R^* is free.   However, the action of R^* on R^2
in Example 2 is not free because every element of R^* has a fixed
point at the origin.

Given a Lie group G acting on the right of a manifold P ,
the <u>orbit of G through</u> $u \in P$ is the set

$$[u]_G = \{R_g(u) \,|\, g \in G\}$$

and we denote by P/G the set of all such orbits for $u \in P$.   This

factor space is topologized by requiring that the projection

$$\Pi_p : P \longrightarrow P/G : u \longmapsto [u]_G$$

is continuous.

We shall denote the connected component of the identity in a Lie group G by G^+ . The topology on G is second countable if and only if the quotient group G/G^+ is countable (or finite). By a Lie subgroup H of a Lie group G we shall mean a subgroup that is also a submanifold and a Lie group with respect to this structure; we shall also require H to be second countable to ensure uniqueness (cf. [57] p.40).

Example 4

When G = $G\ell(1;R)$ = R^* we find

$$G^+ = G\ell(1;R)^+ = R^+ = \{g \in R \mid g > 0\} .$$

An illustrated study of the geometry of the Lie group $S\ell(2;R)$, consisting of unimodular operators on R^2, is given in Dodson and Poston [28]. Lie groups are special cases of topological groups, for a study of the category of the latter see Higgins [48].

Next we collect some important general results.

1.5.1   Theorem

Let G be a locally compact topological group, H a Lie group and h : G $\longrightarrow$ H a continuous homomorphism with h injective on some neighbourhood of the identity in G .

Then G is a lie group.

Proof

Hochschild [50] Ch. 8.   □

Corollary

Every closed subgroup of a Lie group is locally compact and therefore a Lie group.   □

1.5.2   Theorem

Let G be a simply connected, connected Lie group and H a normal, connected Lie subgroup. Then H is closed in G .   Also, the canonical map

$$\pi_H : G \longrightarrow G/H$$

is smooth, and there exists a smooth map

$$\rho_H : G/H \longrightarrow G$$

such that $\pi_H \circ \rho_H = I_{G/H}$ and

$$H \times (G/H) \longrightarrow G : (x,u) \longrightarrow x\rho_H(u)$$

is an isomorphism in Man.

Proof

Hochschild [50] p.135. Note that H and G/H are necessarily simply connected. ☐

1.5.3   Theorem

Let H be a connected Lie subgroup of a connected Lie group G. If the closure in G of every one-dimensional connected Lie subgroup of H lies in H , then H is closed in G .

Proof

Hochschild [50] p.192. ☐

1.5.4   Theorem

Every continuous homomorphism of a connected Lie group into a connected Lie group is smooth.

Proof

Hochschild [50] p.84. ☐

1.5.5   Theorem

Any topological group which has a C^1-Lie group structure admits a unique C^∞-Lie group structure compatible with the given C^1 structure.

Proof

Pontryagin [77]. ☐

Remark

Data on the commoner Lie groups (general linear, special linear, orthogonal, unitary, special orthogonal, special unitary, symplectic and spin) can be found in Porteous [78] p.421 et seq. For more of the general theory see Chevalley [12] and Kobayashi [56], for detailed applications see DeWitt and DeWitt [23].

2.   STRUCTURES ON MANIFOLDS

2.1   *Vector bundle, exact sequence*

We have seen in §1.3 that local tangent vector spaces arise naturally on (smooth) manifolds.   These vector spaces can be fitted together to form a new manifold, the tangent bundle.   More generally, we can use other vector spaces to make arbitrary vector bundles over a given manifold.   Later we obtain tangent bundles via a functor from the category of manifolds to the category of vector bundles.   We outline the construction of the category of vector bundles, following mainly the development in Lang [60] Ch. 2. Throughout, our manifolds and maps will be assumed to be smooth.

A vector bundle with fibre   V   over a manifold   M   is a triple $(M.E,\Pi)$ where   E   is a manifold,   V   is a vector space and

$$\Pi : E \longrightarrow M$$

is a surjection satisfying:

(i) $(\forall x \in M)\, \Pi^{\leftarrow}(x)$ is isomorphic to   V   in Vec.

(ii) $(\forall x \in M)$ $(\exists$ open   U   containing   x) and a diffeomorphism

$$\tau : \Pi^{\leftarrow}(U) \longrightarrow U \times V$$

commuting with the projections

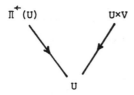

and the induced maps on fibres

$$\tau_x : \Pi^{\leftarrow}(x) \longrightarrow V$$

are isomorphisms in Vec.   (Since we are in finite dimensions the

τ_x are also isomorphisms in Top because the usual topology always makes linear maps continuous.)

We call M the base space, E the total space and Π the bundle map of (M,E,Π). The class of vector bundles becomes a category, V bun, with morphisms defined as follows.

A vector bundle morphism between two vector bundles (M,E,Π), (M',E',Π') is a pair of maps

$$h_o \; : \; M \longrightarrow M' \; , \quad h \; : \; E \longrightarrow E'$$

satisfying:

$$
\begin{array}{ccc}
 & h & \\
E & \longrightarrow & E' \\
\Pi \downarrow & \text{commutes} & \downarrow \Pi' \\
M & \longrightarrow & M' \\
 & h_o &
\end{array}
\qquad
\left\{
\begin{array}{l}
\text{and each restriction} \\[2ex]
h_x \; : \; \Pi^{\leftarrow}(x) \longrightarrow \Pi'^{\leftarrow}\!\big(h(x)\big) \\[2ex]
\text{is a continuous linear map.}
\end{array}
\right.
$$

In applications we often use several vector bundles over a given manifold; then h_o is the identity.

In Lang [60] will be found fuller descriptions, extended to the case of infinite dimensional manifolds with vector bundles modelled on Banach spaces, and details of operations on vector bundles. We note that unlike Man, the category V bun does admit direct sums (the Whitney sum), quotients and pullbacks (cf. Hirsch [49] for specific constructions). Functional operations available in Vec such as tensor products and exact sequences can also be carried over to V bun (cf. Lang [60] Ch. 3). Vector bundles over a given manifold M form a subcategory V bun(M) (cf. II §1.5). There is a zero object in V bun(M) (cf. II §1.9), the zero-dimensional bundle over M, denoted {O}. We shall look next at exact sequences, but first we point out that these, like the constructions above, will be for smooth vector bundles over smooth manifolds. The whole thing could be presented for C^r-bundles over C^k-manifolds but then care is necessary when differential processes arise and certain operations cease to be closed. In any particular application it is not difficult to make the appropriate substitution of known classes of differentiability;   it is tedious to do so generally.

The idea of exact sequences (cf. §1.1) can be transferred to vector bundles (cf. Lang [60] p. 48 and Hirsch [49] p. 93). Given a manifold M and two vector bundles (M,E,Π), (M,E',Π') over M with fibres V and V' respectively then a diagram (in V bun)

$$\{\underline{o}\} \longrightarrow E' \xrightarrow{\ f\ } E$$

is called an __exact sequence__ if f is injective and M has an open cover K such that for each $U\epsilon K$ there is a commutative diagram in Man:

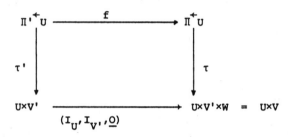

where the τ' and τ are diffeomorphisms.

Now we can define $S\subset E$ to determine a __subbundle__ of (M,E,Π) if there is an exact

$$\{\underline{o}\} \longrightarrow E' \xrightarrow{\ f\ } E$$

with $f(E')$ = S   (cf. Examples in §1.1).

Similarly, a diagram (in V bun)

$$E \xrightarrow{\ g\ } E'' \longrightarrow \{\underline{o}\}$$

is an __exact sequence__ if g is surjective and M has an open cover K such that for each $U\epsilon K$ there is a commutative diagram in Man:

$$
\begin{array}{ccc}
\Pi^{\leftarrow}U & \xrightarrow{\quad g \quad} & \Pi''^{\leftarrow}U \\
\Big\downarrow{\tau} & & \Big\downarrow{\tau''} \\
U\times V = U\times V'\times W & \xrightarrow{(I_U, \Pi_W)} & U\times W
\end{array}
$$

where the τ and τ'' are diffeomorphisms.

We can tie these definitions in with those in Vec (cf. §1.1)
by defining for any vector bundle morphism (cf. §2.1)

$$
\begin{array}{ccc}
E & \xrightarrow{\ h\ } & E' \\
\Pi \downarrow & & \downarrow \Pi' \\
M & \xrightarrow[\ h_o\ ]{} & M'
\end{array}
\qquad
\left\{
\begin{array}{l}
(\forall x \in M)\ \text{the restriction} \\[2mm]
h_x : \Pi^{\leftarrow}(x) \longrightarrow \Pi'^{\leftarrow}(h(x)) \\[2mm]
\text{is linear.}
\end{array}
\right.
$$

$$
\ker(h) \;=\; \bigcup_{x \in M} \ker(h_x) \qquad \text{(the \underline{kernel} of h)}
$$

$$
\mathrm{im}(h) \;=\; \bigcup_{x \in M} \mathrm{im}(h_x) \qquad \text{(the \underline{image} of h)} .
$$

The kernel and image of a vector bundle morphism naturally inherit
vector bundle structures because kernels and images of linear maps
are necessarily vector subspaces (cf. Lang [60] §3 for further
details).   Now we can say that a diagram in V bun of the form

$$
\cdots E_{-1} \xrightarrow[f_{-1}]{\ } E_o \xrightarrow[f_o]{\ } E_1 \xrightarrow[f_1]{\ } E_2 \cdots
$$

is called an **exact sequence** if

$$
(\forall k)\ (\forall x \in E_k)\ \mathrm{im}\, f_{kx} \;=\; \ker f_{k+1}{}_{f_k(x)} .
$$

Then a **short exact sequence** is one of the form

$$
\{\underline{o}\} \longrightarrow E' \xrightarrow{\ f\ } E \xrightarrow{\ g\ } E'' \longrightarrow \{\underline{o}\}
$$

and then f is a monomorphism, g is an epimorphism (cf. II
§1.8) and im f = ker g.   Such a short exact sequence is called
split if there is a monomorphism h : E'' $\longrightarrow$ E   such that
g∘h = $I_{E''}$ (cf. II §1.10).   As before (cf. §1.1) we can view E''
in the short exact sequence as a quotient object of the monomorphism
f , this **quotient bundle** is as usual determined up to isomorphism.
In the particular case that E'⊂E is a sub bundle then the
quotient bundle is denoted E/E' and its fibres are the quotient
vector spaces $\Pi^{\leftarrow}(x)/\Pi'^{\leftarrow}(x)$.

We pointed out earlier that V bun admits pullbacks (cf. II

§2.2).   Given a vector bundle (M,E,Π) and a manifold map
$f : M_o \longrightarrow M$ the _induced bundle_ or _pullback_ over M_o is

$$(M_o, f^*E, \Pi_o) \text{ with } f^*E = \{(x,y) \in M_o \times E \mid f(x) = \Pi(y)\}$$

and

$$\Pi_o : f^*E \longrightarrow M_o : (x,y) \longmapsto x .$$

The atlas for f^*E is the maximal one containing all charts

$$\psi : \overleftarrow{f}U \longrightarrow R^n$$

for which there is a chart on E ,

$$\phi : U \longrightarrow R^n$$

satisfying

$$z \in \overleftarrow{f}U, \ f(z) = x \in U \implies \psi|_{\overleftarrow{\Pi_o}(z)} = \phi|_{\overleftarrow{\Pi}(x)} .$$

2.1.1   **Theorem**   (Covering homotopy theorem)

Let B be paracompact and suppose that E is a vector bundle over
$B \times [0,1]$.   Then E is isomorphic to the vector bundle $E_B \times [0,1]$
where:

$$E_B = E|_{B \times \{o\}} .$$

Proof

Hirsch [49] p. 90.   □

2.1.2   **Theorem**

Let B be paracompact, $f,g : B \longrightarrow M$ homotopic maps and let
(M,E,Π) be a vector bundle.   Then f^*E is isomorphic to g^*E
in V bun .   In particular if g is constant then f^*E is
trivial.

Corollary

Every vector bundle over a contractible paracompact space is
trivial.

Proof

Hirsch [49] p. 97.   (See III §1.7 for the definition of homotopy
and contractibility.)   □

We conclude this section with some applications of the above (cf. Hirsch [49] p. 94 et seq.).   The <u>Whitney sum</u> of bundles E',E over M is the bundle E' ⊕ E with fibres, $\Pi'^{\leftarrow}(x) \oplus \Pi^{\leftarrow}(x)$ for x∈M .   Direct sums of charts for E',E give charts for E' ⊕ E and the exact sequences of fibres (in Vec)

$$\{\underline{0}\} \longrightarrow \Pi'^{\leftarrow}(x) \longrightarrow \Pi'^{\leftarrow}(x) \oplus \Pi^{\leftarrow}(x) \longrightarrow \Pi^{\leftarrow}(x) \longrightarrow \{\underline{0}\}$$

can be collected to give a (split) exact sequence (in V bun)

$$\{\underline{0}\} \longrightarrow E' \longrightarrow E' \oplus E \longrightarrow E \longrightarrow \{\underline{0}\} .$$

2.1.3   Theorem

Every short exact sequence of vector bundles over a paracompact manifold M is split.

Proof

We seek a left inverse (cf. II §1.10) of a monomorphism f : E' ⟶ E .   Locally this is available since f(E') is a subbundle of E and there are open U⊆M with charts

$$\Pi'^{\leftarrow}U \longrightarrow U{\times}R^n \longrightarrow R^m{\times}R^n$$
$$\Pi^{\leftarrow}U \longrightarrow U{\times}R^k \longrightarrow R^m{\times}R^k .$$

The inverse then appears as a linear map $R^n \longrightarrow R^k$.   These local left inverses are then glued together by a partition of unity, available on M by its paracompactness (cf. III §1.6).   □

2.1.4   Theorem

For every vector bundle E there exists a vector bundle E' , over the same paracompact base M , such that E ⊕ E' is trivial.

Proof

The technique is again to use a partition of unity and to construct E ⊕ E' = M${\times}R^m$, where m is the (finite) number of generators of the module of sections of the bundle E .   Details will be found in Greub, Halperin and Vanstone [40] p. 76 et seq.   □

2.1.5   Theorem

A vector bundle E over M , with fibre V and dim V = k , is trivial if and only if there exist k sections

$$\sigma_i \; : \; M \longrightarrow E \; : \; x \longmapsto \sigma_i(x)$$

such that ($\forall x \in M$)

$$\{\sigma_i(x) \mid i = 1,2,\ldots,k\} \quad \text{is linearly independent.}$$

Proof

Such a family of sections gives an isomorphism with the trivial
bundle $M \times \mathbb{R}^k$. $\square$ (Cf. also Theorem 3 in §2.3 below.)

2.2   *Tangent bundle, differential, jet*

The tangent bundle functor T is a covariant functor from Man to
V bun defined by

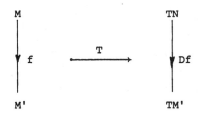

where (cf. II §2.2):

(i) $TM = (M, \; \coprod_{x \in M} T_x M, \; \Pi_T)$.

(ii) Π_T sends each $y \in T_x M$ to x ,

(iii) the commuting diagram for Df is

$$
\begin{array}{ccc}
TM & \xrightarrow{\;\;Df\;\;} & TM' \\
\Pi_T \downarrow & & \downarrow \Pi'_T \\
M & \xrightarrow{\;\;f\;\;} & M'
\end{array}
\qquad
\left\{ \; \text{and } Df\big|_{T_x M} = D_x f \; . \right.
$$

Thus, if M is an n-manifold then TM is a vector bundle
with fibre $\mathbb{R}^n$. The bundle morphism Df is called the
differential of f .   Some authors denote Df by f_* .   For full
details of the synthesis of the differentiable structure on TM

from the atlas for M see Dodson and Poston [28].

The results of the previous section on vector bundles have various particular applications to the tangent bundle.   Firstly, from Lang [60] p. 53, given a manifold immersion (cf. §1.4)

$$f : M \longrightarrow M' \qquad \text{(in Man)}$$

we have a differential (in fact a monomorphism)

$$Df : TM \longrightarrow TM' \qquad \text{(in Man)}$$

but also a canonical vector bundle morphism

$$T*f : TM \longrightarrow f*TM', \quad \text{(in V bun)}$$

by the universal property of pullbacks (cf. II §2.3).
Then there is an exact sequence

$$\{o\} \longrightarrow TM \xrightarrow{T*f} f*TM' .$$

We can also work with submersions.   In particular for a vector bundle (M,E,Π) we obtain the exact sequence (for $D\Pi$ is epic here)

$$\{o\} \longrightarrow \Pi*E \longrightarrow TE \longrightarrow \Pi*TM \longrightarrow \{o\} .$$

Next, from Hirsch [49] p. 94, given a vector bundle (M,E,Π) we can identify each vector space $\Pi^{\leftarrow}(x)$ with its own tangent space $T_x \Pi^{\leftarrow}(x)$.

Hence, E is a sub bundle of $T_M E$, the restriction of TE to M⊂E , well defined since M is isomorphic to the zero section of E .   But then TM is also a sub bundle of $T_M E$ via the differential of the inclusion $i : M \hookrightarrow E$.   It follows that there is a short exact sequence

$$\{o\} \longrightarrow E \longrightarrow T_M E \xrightarrow{D\Pi} TM \longrightarrow \{o\}$$

which is split by

$$Di : TM \longrightarrow T_M E .$$

In consequence there is a natural isomorphism

$$T_M E \equiv E \oplus TM$$

with the Whitney sum (cf. §2.1).

There are many ways to view tangent vectors on a manifold: equivalence classes of (C^1) curves going through a given point

in a given direction; derivations on the algebra of (smooth) real
functions on the manifold; equivalence classes of (C^1) real
functions agreeing up to the first derivative; as geometrical
directions in R^{2n}, by virtue of the Whitney embedding theorem
for manifolds with dimension n (and second countable topology,
cf. III §1.5).

It is the aspect of 'differential agreement up to order one'
that suggests the generalisation to <u>jets</u>. A tangent vector is a
<u>1-jet</u> or a <u>jet of order 1</u>; we proceed to construct n-jets.
Denote by $T^o M$ the (infinite-dimensional) vector space of smooth
real functions on a manifold M .   (The notation is chosen to be
consistent with that for general tensor fields in the next section.)
Now we follow Schwartz [87] p.2 and define:

> $g \in T^o M$ is <u>n-horizontal</u> at $x \in M$ if, in the
> co-ordinates of some (hence any) chart, all
> partial derivatives of g up to and including
> order n vanish at x .

Then an n-jet to M at x is a linear functional $\tau \in (T^o M)^*$ such
that

$$g \in T^o M \text{ is n-horizontal} \implies \tau(g) = 0 .$$

Now we find local co-ordinate expressions for n-jets .

Let

$$\phi : U \longrightarrow R^n : \alpha \longmapsto (\alpha^1, \alpha^2, \ldots, \alpha^n)$$

be a chart about $x \in M$.   Hence any $g \in T^o M$ has a local expression

$$g|_U \circ \overset{\leftarrow}{\phi} : R^n \longrightarrow R : (\alpha^i) \longmapsto \tilde{g}(\alpha^i) = g(\alpha)$$

and, in fact near $\underline{0} \in R^n$,

$$\tilde{g}(\alpha^i) = C + \rho(\alpha^i) + H_n(\alpha^i)$$

where C is some constant, $H_n(\alpha^i)$ is the <u>n-horizontal part</u> and
$\rho(\alpha^i)$ is the so-called <u>principal part</u>.   For any n-jet τ at
$x \in M$, by construction,

$$\tau(H_n(x^i)) = 0 = \tau(C)$$

and so, by the linearity of τ

$$\tau\left(\tilde{g}(x^i)\right) = \tau\left(\rho(x^i)\right)$$

$$= \tau\left(\sum \partial_i \tilde{g}\big|_o x^i + \cdots \frac{1}{n!} \sum \partial_{i_1} \cdots \partial_{i_n} \tilde{g}\big|_o x^{i_1} \cdots x^{i_n}\right)$$

$$= \sum \partial_i \tilde{g}\big|_o \tau(x^i) + \cdots \frac{1}{n!} \sum (\partial_{i_1} \cdots \partial_{i_n} \tilde{g}\big|_o) \tau(x^{i_1} \cdots x^{i_n}).$$

Now we define the co-ordinates of τ by

$$\tau(x^i) = \tau^i, \ldots, \tau(x^{i_1} \cdots x^{i_n}) = \tau^{i_1 \cdots i_n}.$$

Hence a chart about $x \in M$ yields a chart about τ.
So there is an unambiguous representation

$$\tau(g) = \sum \partial_i \tilde{g}\big|_o \tau^i + \cdots + \frac{1}{n!} \sum \partial_{i_1 \cdots i_n} \tilde{g}\big|_o \tau^{i_1 \cdots i_n}.$$

In each term the summation is from $i=1$ to $i=\dim M$. If we use
the summation convention and write the co-ordinates before the
derivatives, omitting the function g we have

$$\tau = \tau^i \partial_i + \cdots + \frac{1}{n!} \tau^{i_1 \cdots i_n} \partial_{i_1 \cdots i_n}.$$

Then we easily recognise the case of the 1-jet as a tangent vector

$$\tau = \tau^i \partial_i \in T_x M,$$

that is a derivation on smooth real functions.

Evidently the n-jets at $x \in M$ admit finite linear combinations
over R and so we have the n-jet vector space $J_x^n M$ to M at x.
These spaces collect into a vector bundle (cf. Palais [74]), glued
together via the above charts and those on M, the n-jet bundle
$J^n M$. Sections of $J^n M$ yield n-jet fields and these with
pointwise addition form an (infinite-dimensional) vector space in
a similar way to the tensor fields. Moreover, there is a natural
composition of jet fields arising from the natural composition of
partial derivations in coordinates. So if τ is an n-jet field
and σ is an m-jet field then $\tau\sigma$ is an (n+m)-jet field. For
more details of the theory see Yano and Ishihara [107] where
chapter X is particularly useful for 2-jet bundles.

(We note in passing that the manifold structure on $J^n M$ is of class C^k if M is of class C^{n+k} . Another generalisation, discussed in Hirsch [49], is the notion of n-jets from M to N where N is some manifold. These are equivalence classes of maps M $\longrightarrow$ N agreeing up to their n-th derivative.)

An integral curve of a (tangent) vector field w on M is a curve c in M such that (cf. §1.3)

$$(\forall t \in \text{dom } c) \quad \dot{c}(t) = w \circ c(t) \ .$$

Such curves always exist for smooth vector fields and they are essentially unique. (cf. Dodson and Poston [28] Appendix for a recent geometric proof.) Intuitively such curves are found by 'joining up the arrows' given by the vector field as a tangent vector at each point. The process allows a generalization to 'joining up subspaces' in the following way (cf. Kobayashi and Nomizu [57] p. 10, Brickell and Clark [9] ch. 11).

A distribution S of dimension r on M is an assignment to each x∈M of an r-dimensional subspace S_x of $T_x M$. We shall consider only those distributions that are differentiable, that is every x∈M has a neighbourhood U on which there are r vector fields forming a basis for S_y , y∈U . So we can think of S as a 'field of tangent subspaces'. A vector field w on M is said to belong to S if

$$(\forall x \in M) \quad w(x) \in S_x \ .$$

A distribution S is called involutive if : whenever w,v belong to S then so does [w,v], and here [,] is the Lie bracket or commutator :

$$[w,v]f = w(v(f)) - v(w(f)) \qquad (\forall f \in T^o M) \ .$$

A connected submanifold N of M is called an integral manifold of a distribution S if

$$(\forall x \in N) \quad D_x i \ (T_x N) = S_{i(x)}$$

where i is the inclusion N $\hookrightarrow$ M . Such an N is called maximal if no other integral manifold of S contains it. The generalization of the integral curve result is due to Frobenius (cf. Chevalley [12] for proof):

2.2.1   Theorem

Let S be an involutive distribution on a manifold M .   Through
every x∈M there passes a unique maximal integral manifold N(x)
of S .   Any integral manifold through x is an open submanifold
of N(x) .   □

 We shall make use of distributions in two contexts; one
arises from a connection (cf. §2.6 below) and the other arises
from a Lie algebra (cf. §2.5 below).   In the latter case the
distribution is naturally involutive and indeed these play an
important role in modern geometric quantization theory (cf. e.g.
Simms and Woodhouse [89]).   Another result is the following, also
proved in Chevalley [12] (cf. [57] p. 11).

Proposition

Let S be an involutive distribution on a manifold M .   Let N
be a submanifold of M whose connected components are all integral
manifolds of S .   Given a manifold map f : M' ⟶ M with
fM'⊆N then f is a manifold map from M' into N if N is
second countable.   □

2.3   Fibre bundle : principle, frame, associated.

A useful survey on fibre bundles is given in Eells [31].
 We saw that vector bundles arise naturally over manifolds.
More generally, we have the concept of a bundle (B,X,ρ) where

$$\rho : B \longrightarrow X \quad \text{(in Top)}$$

is a continuous surjection and B is viewed as a union of fibres
$\{\rho^{\leftarrow}(x) \mid x \in X\}$ glued together via the topology of X .   In our
applications we usually have more structure for the fibres, usually
of an algebraic character, these are the fibre bundles resulting
from actions of groups.   The simplest case is of vector bundles,
where the fibre algebra is the additive group F^n for some n .
Next, we wish to admit more general, not necessarily Abelian, group
structure for fibres.   However, we do not wish to lose the
compatibility of the algebra with the topology so we need
topological groups and in the case that our spaces are manifolds

we shall demand compatibility of the algebra with the differential
structure so we need <u>Lie groups</u> (cf. §1.5).   Many of the bundles
we shall use actually arise as associated bundles of a bundle of
the following type.

A <u>principal fibre bundle</u> over a manifold M is a triple
(P,G,M) where P is a manifold and G is a Lie group such
that:

(i)   G acts freely on P to the right,

(ii)   M = P/G and the canonical projection

$$\Pi_P : P \longrightarrow M : u \longmapsto [u]$$

is smooth,

(iii) P is <u>locally trivial</u>; that is, every $x\epsilon M$ has a neighbourhood
U such that $\overset{\leftarrow}{\Pi_P}U$ is diffeomorphic to U×G .

We call G the <u>structure group</u> of the bundle; property (ii)
ensures a transitive action on all fibres $\overset{\leftarrow}{\Pi_P}(x)$.

The <u>vertical subspace</u> of the tangent space T_uP for $u\epsilon P$ is
the kernel of $D\Pi_P$:

$$G_u = \{X\epsilon T_uP | D\Pi_P X = 0\epsilon T_{\Pi_P(u)}M\} .$$

<u>Example 1</u>

The trivial product bundle,   P = M×G .

<u>Example 2</u>

The universal covering manifold $\tilde{M}$ (cf. §1.2) is a principal fibre
bundle over M with structure group $\pi_1(M)$, the <u>fundamental
group</u> of M (cf. Singer and Thorpe [90] and ch. III §1.7).   Note
that Chevalley [12] called this group the <u>Poincaré group</u>.

<u>Example 3</u>

The <u>frame bundle</u> over a manifold M :

$$P = LM = \{(x,(X_i)) | x\epsilon M; (X_i) \text{ an ordered basis for } T_xM\}$$
$$G = G\ell(n;\mathbb{R}) ,$$

where n is the dimension of M .
This example is important in applications for we can derive from it

all of the <u>tensor bundles</u>.   We see that given any basis for T_xM
at $x\in M$ (e.g. by choosing a chart about x) we obtain an
isomorphism

$$T_xM \longrightarrow R^n : \alpha \longmapsto (\alpha^i)$$

which allows $G\ell(n;R)$ to act by matrix multiplication on the
coordinates (α^i) .   Hence the orbit through $(x,(X_i))\in LM$ is
given by

$$[(x,(X_i))] = \{(x,(X_j g_i^j)) \mid (g_i^j) = g\in G\ell(n;R)\} .$$

But as g runs through $G\ell(n;R)$, so $R_g(x,(X_i))$ runs through all
bases for T_xM .   Hence, we may as well abbreviate $[(x,(X_i))]$ to
x , so obtaining the projection map

$$\Pi_L : LM \longrightarrow M : (x,(X_i)) \longmapsto x .$$

We can see how LM and TM are related by the tangent bundle
functor (cf. §2.2) as follows:
The canonical projections

$$\Pi_T : TM \longrightarrow M, \Pi_L : LM \longrightarrow M$$

yield an exact sequence (cf. §2.1)

$$\{\underline{0}\} \longrightarrow \text{kernel } (T\circ\Pi_L) \longrightarrow TLM \longrightarrow \Pi_L^*(TM) \longrightarrow \{\underline{0}\}$$

and it follows that kernel $(T\circ\Pi_L)$ and the pullback (cf. §2.1)
$\Pi_L^*(TM)$ are trivial vector bundles.

The <u>canonical one-form</u> of a frame bundle LM is the map

$$\theta : TLM \longrightarrow R^n$$

$$: (u,W) \longmapsto \Pi_u\circ D\Pi_L(u,W)$$

where for all $u = (x,(u_j))\in LM$

$$\Pi_u : T_xM \longrightarrow R^n$$

$$: \alpha^i u_i \longmapsto (\alpha^i) .$$

We note that, by construction, vertical vectors lie in the kernel
of $D\Pi_L$ so $\theta(G_u) = \underline{0}$ for all u .   In components we see that θ
is a matrix action:

$$\theta \; : \; TLM \; \longrightarrow \; R^n$$

$$: \; (x^i, b^i_j, x^i, B^i_j) \; \longrightarrow \; [b^i_j]^{-1} \; [x^i] \; .$$

Here, $(x^i) \in R^n$ are coordinates of $x \in M$ with respect to some chart (U, ϕ) and these determine basis fields $(\partial_i)_{i=1,\ldots,n}$ with respect to which a frame u is given by

$$(b^i_j \partial_i)_{j=1,\ldots,n} \; = \; (u_j)_{j=1,\ldots,n} \; .$$

The x^i are components with respect to $(\partial_i)_{i=1,\ldots,n}$ of a typical $X \in T_x M$. Evidently we have, using the summation convention,

$$X = x^i \partial_i = Y^j u_j = Y^j b^i_j \partial_i$$

for some $[Y^j] = [b^i_j]^{-1}[x^i] \in R^n$. So θ gives the components of X with respect to the frame u .

Example 4

The trivial product bundle over S^1, $LS^1 = S^1 \times R^*$ has two components, corresponding to positively oriented bases

$$L^+ S^1 \; = \; S^1 \times R^+$$

and to negatively oriented bases

$$L^- S^1 \; = \; S^1 \times R^- \; .$$

Quite generally, LM has two components if M is orientable otherwise LM is connected. Precisely, a manifold M with atlas $\{(U_\alpha, \phi_\alpha) \, | \, \alpha \in A\}$ is orientable if there is a sub-atlas $\{(U_\alpha, \phi_\alpha) \, | \, \alpha \in A'\}$ such that whenever $U_\alpha \cap U_\beta \neq \emptyset$ for $\alpha, \beta \in A'$ then the change of chart map $\phi_\alpha \circ \phi_\beta^{\leftarrow}$ has a Jacobian matrix with positive determinant.

Example 5

Any manifold admitting a two-chart atlas is orientable if the intersection of the chart domains is connected. For the Jacobian, representing a diffeomorphism between subsets of R^n , cannot have zero determinant. It follows that all of the spheres S^n are orientable because each admits a two-chart atlas : stereographic projection from its north and south poles (cf. Porteous [78] p. 169).

Given a principal fibre bundle (P,G,M) and a manifold F on which G acts on the left, then the <u>fibre bundle associated to (P,G,M) with fibre F</u> is a manifold $(P{\times}F)/G$ defined by

(i)   the right action of G on $P{\times}F$ is

$$P{\times}F{\times}G \longrightarrow P{\times}F : (u,a,g) \longmapsto (R_g(u),L_{g^{-1}}(a)) ,$$

(ii) the projection map is

$$\Pi_F : (P{\times}F)/G \longrightarrow M : R_G(u,a) \longmapsto \Pi_P(u)$$

and it is required to be smooth,

(iii) every $x{\in}M$ has a neighbourhood U such that $\overset{\leftarrow}{\Pi}_F U$ is diffeomorphic to $U{\times}F$.

<u>Example 6</u>

Take $F = G$ and use left translation.

<u>Example 7</u>

The tangent bundle TM is the bundle associated to LM with fibre R^n , where $n = \dim M$. We can identify each fibre $\overset{\leftarrow}{\Pi}_R(x)$ with the tangent space $T_x M = \overset{\leftarrow}{\Pi}_T(x)$ (cf. §2.1). In particular for S^1 with $F = R^1 = R$, $G = R^*$ we have

$$(LS^1{\times}R){\times}R^* \longrightarrow LS^1{\times}R$$
$$: (x,b,a,g) \longmapsto (x,bg,g^{-1}a) ,$$

for the right action of R^* on $LS^1{\times}R$. Hence the orbit of R^* through $(x,b,a){\in}LS^1{\times}R$ is

$$[(x,b,a)] = \{(x,bg,g^{-1}a) \,|\, g{\in}R^*\}$$

which we identify with $(x,a){\in}T_x S^1$ because bg runs through all bases for $T_x S^1$ while g runs through R^* .

The tangent bundle is a special case of the <u>tensor bundles</u> $T^k_h M$ for $k,h = 0,1,\ldots$. These are associated to the frame bundle LM but here F is a tensor product of k copies of R^n and h copies of its dual R^{n^*} . The general linear group acts independently on the factors of the tensor product, as for TM on R^n and via transposed inverses on R^{n^*} (the contravariant nature of dual taking, again). We identify $T^0_0 M$ with $M{\times}R$, $T^1_0 M$ with

TM, and $T_1^O M$ with TM^* the <u>cotangent vector bundle</u> with fibre
$\mathcal{R}^{n*}$. When necessary we shall denote the projection maps onto M
by

$$\Pi_h^k : T_h^k M \longrightarrow M .$$

To be more categorical we could introduce a family of functors
from Man to V bun , with images these tensor bundles. Such a
procedure is outlined in Lang [60] p. 57.

From the tensor bundles we can define the <u>tensor fields</u>,
essential in the formulation of physical theories and in the
geometrical study of manifolds.

A <u>tensor field</u> of type $\binom{k}{h}$ is a (smooth) section of Π_h^k
(cf. II §1.10). That is some

$$W : M \longrightarrow T_h^k M \qquad \text{(in Man)}$$

such that $\Pi_h^k \circ w = I_M$.

The set of such fields forms a vector space, $T_h^k M$, with
pointwise addition and in particular there always exists the <u>zero</u>
<u>section field</u> of any type.

A <u>morphism of principle fibre bundles</u> is defined analogously
to vector bundle morphisms (cf. §2.1); instead of linear maps on
the fibres we need group homomorphisms. A subset (P',G',M) of
(P,G,M) is a <u>sub-bundle</u> if the inclusions form a principle fibre
bundle morphism; in this case we call (P',G',M) a <u>reduced bundle</u>
and we say that G is reducible to G' .

2.3.1   <u>Theorem</u>

The structure group G of a principal fibre bundle (P,G,M) is
reducible to a Lie subgroup G' if and only if there is an open
cover $\{U_\alpha | \alpha \epsilon K\}$ of M and a set of maps (<u>transition functions</u>)

$$\psi_{\beta\alpha} : U_\alpha \cap U_\beta \longrightarrow G'$$

satisfying for all $x \epsilon U_\alpha \cap U_\beta \cap U_\gamma$

$$\psi_{\beta\alpha}(x) = \psi_{\gamma\beta}(x) . \psi_{\beta\alpha}(x) .$$

Note that G' need not be closed in G (cf. Theorem 2 below).

Proof

Kobayashi and Nomizu [57] p. 51-54.   □

Remark

The locally trivial structure of a principal fibre bundle   (P,G,M)
always provides transition functions taking values in   G.   For
there is an open cover $\{U_\alpha | \alpha \epsilon K\}$ of   M   with diffeomorphisms

$$\psi_\alpha : \overleftarrow{\Pi_P} U_\alpha \longrightarrow U_\alpha \times G : u \longmapsto (\Pi_P(u), \phi_\alpha(u))$$

satisfying $\phi_\alpha \circ R_g(u) = R_g \circ \phi_\alpha(u)$.

Hence if $u \epsilon \overleftarrow{\Pi_P} (U_\alpha \cap U_\beta)$

$$\left(\phi_\beta \circ R_g(u)\right) \left(\phi_\alpha \circ R_g(u)\right)^{-1} = \phi_\beta(u) \left(\phi_\alpha(u)\right)^{-1} ,$$

depending on $\Pi_P(u)$ only, not on   u .

The required functions are given by

$$\psi_{\beta\alpha} \circ \Pi_P(u) = \phi_\beta(u) \left(\phi_\alpha(u)\right)^{-1} .$$

Kobayashi and Nomizu [57] p. 52 show how a principle fibre
bundle can be constructed over   M ,   given some transition
functions taking values in a Lie group   G .   We give an example
of the construction of a reduced bundle of LS^1,   with structure
group   Z   in §2.6 below.

2.3.2   Theorem

The structure group   G   of a principal fibre bundle   (P,G,M)   is
reducible to a closed subgroup   G'   if and only if the associated
bundle with fibre   G/G'   admits a section

$$\sigma : M \longrightarrow E = P/G' .$$

Proof

Kobayashi and Nomizu [57] p. 57.   □   (Cf. §2.8 Theorem 5.)

2.3.3   Theorem

A principal fibre bundle   (P,G,M)   is trivial, that is   P = M×G ,
if and only if   P   admits a section $\sigma : M \longrightarrow P$.

Proof

Steenrod [96] p. 36.   (Cf. also the remark on flat connections at the end of §2.6.)   □

Corollary

(P,G,M) is trivial if and only if it is isomorphic to the induced
bundle f*Q where Q is the product bundle over a point q and
f is the unique constant map.

Proof

The induced bundle, or pullback over M (cf. §2.1) of a constant
map $f : M \longrightarrow \{q\}$ is (f*Q,G,M) where

$$f*Q = \{(x,u) \in M \times P \mid f(x) = \Pi_p(u)\}$$
$$= \{(x,u) \in M \times \Pi_p^{\leftarrow}(q)\}$$

and the canonical projection is

$$(x,u) \longmapsto x .$$

Induced bundles of trivial bundles are trivial and the existence
of the above isomorphism is equivalent to the existence of a
section $\sigma : M \longrightarrow P$.   Details are given in Husemoller [52]
p. 48, together with further applications of induced bundles.   □

Remark

A **sufficient** condition for (P,G,M) to be trivial is for M (**or**
the fibres of P) to be contractible (cf. III §1.7).   In
particular this will happen if M is homeomorphic to R^n (**or** if
each $\Pi_p^{\leftarrow}(x)$ is homeomorphic to R^m , $\forall x \in M$).   (Cf. also Theorem 5
in §2.1.)

2.4   *Parallelization*

We can always find smooth sections of vector bundles, that is
vector fields, but this is not in general true for principal fibre
bundles.   In particular we shall have an interest in the frame
bundle LM (cf. §2.3) and the only compact two-manifolds having a
continuous section of their frame bundle are the Klein bottle and
torus while the only spheres with this property are S^1, S^3 and
S^7 (cf. Brickell and Clark [9] §7.3, Schwartz [87] ch. 10 and
Husemoller [52] ch. 15).   The difference between finding sections

of the vector bundle TM and the fibre bundle LM is that we
need n = dim M linearly independent sections of TM to give a
section of LM . In the case of S^2 , the Hairy Ball Theorem
asserts that there is not even a continuous section of TS^2 that
is never zero.

A manifold M is called parallelizable if LM admits a
continuous section; if it exists, such a section is called a
parallelization of M . (Note another usage in Hirsch [49] p. 88.)

Manifolds always have local parallelizations about any point
because a chart (U,ϕ) there provides a smooth choice of basis for
tangent spaces about the point. In consequence of this there is
always a local diffeomorphism $\Pi_L^\leftarrow U \equiv U \times G$. Any manifold with a
global chart is inevitably parallelizable; for example $\mathbb{R}^n$ admits
the identity chart. If a manifold M is parallelizable then TM
is globally homeomorphic to $M \times \mathbb{R}^n$, and if the parallelization is
a C^k-section of LM then this homeomorphism is a C^k-diffeomorphism.
Thus, LM is trivial if and only if M is parallelizable.

We have seen how the frame and tangent bundles are related
functorially by the tangent bundle functor (cf. §2.2) which gives
an exact sequence (cf. §2.3)

$$\{\underline{0}\} \longrightarrow \ker(T \circ \Pi_L) \longrightarrow TLM \longrightarrow \Pi_L^*(TM) \longrightarrow \{\underline{0}\} \ .$$

In the case that M is paracompact this sequence splits and LM is
parallelizable (cf. Hirsch [49] p. 98). We shall see later that
this is related to the idea of a connection (cf. §2.6).

We collect here some results associated with parallelizations
(cf. Brickell and Clark [9], Husemoller [52], Kobayashi and Nomizu
[57], Hirsch [49]):

(i)   If M is compact and connected then TM admits a nowhere-
       zero section if and only if the Euler characteristic $\chi(M)$
       is zero. Since

$$\chi(S^n) = 1 + (-1)^n \ ,$$

       only odd-dimensional spheres admit global nowhere-zero vector
       fields. Also, any compact odd-dimensional manifold M has
       $\chi(M) = 0$, so does any closed orientable manifold with a
       nowhere-zero tangent vector field.

(ii)   Any non-compact paracompact manifold admits a nowhere-zero
        vector field.

(iii)   If M admits a nowhere-zero vector field then so does   M×N
        for any manifold N .

(iv)   If M and N are parallelizable then so is M×N .

(v)   Given a parallelization

$$p : M \longrightarrow LM : x \longmapsto (p_i(x))$$

        then the set of all smooth maps

$$f : M \longrightarrow M$$

        satisfying

$$Df \circ p_i = p_i \circ Df , \qquad (\forall i = 1,2,\ldots, \dim M)$$

        forms a group, the automorphism group of p .

        The group of automorphisms of a parallelization for a
        connected manifold is either a discrete group (orbits are
        discrete sets of points) or it can be given the structure
        of a Lie group, with a free action on M (cf. §1.5).   These
        automorphisms are always isometries with respect to the
        standard metric structure induced by a parallelization
        (cf. §2.8 below).

(vi)   Every parallelizable manifold is orientable.

(vii)   Every parallelizable manifold is metrizable and therefore
        paracompact.

(viii) Every Lie group is parallelizable.

(ix)   Any orientable, compact 3-manifold is parallelizable (in
        fact, the compactness can be relaxed, cf. Steenrod [96] p.203
        and p.221.)

Example

A non-trivial parallelization of $\mathbb{R}^2$ is given by

$$p : \mathbb{R}^2 \longrightarrow L\mathbb{R}^2 : (x,y) \longmapsto (e^x \partial_1, e^x \partial_2).$$

In §2.8 it will be used to illustrate that a parallelization

always determines a metric structure and a connection (cf. §2.6)
that is compatible with this structure (sizes of vectors are
unaltered by parallel transport) but the connection need not be
symmetric.

2.5   *Lie algebra*

A **Lie algebra** is a vector space ξ together with a product
structure

$$\xi \times \xi \longrightarrow \xi : (\alpha,\beta) \longmapsto [\alpha,\beta]$$

satisfying

(i)   bilinearity

(ii) $(\forall \alpha \epsilon \xi)\ [\alpha,\alpha]\ =\ 0$

(iii) the **Jacobi identity** $(\forall \alpha,\beta,\gamma \epsilon \xi)$

$$[[\alpha,\beta],\gamma] + [[\beta,\gamma],\alpha] + [[\gamma,\alpha],\beta]\ =\ 0\ .$$

The second part of Hochschild [50] is devoted to a development
of the theory of Lie algebras.   We shall be concerned with
algebras over R , but complex numbers could equally well be used
and the complex analytic Lie groups and their algebras are also
studied in [50] ch. 17.   In fact, it is the natural generation of
Lie algebras from Lie groups that makes them so important:

2.5.1   Theorem

If ξ is a finite-dimensional Lie algebra over R then there
exists a connected Lie group G whose Lie algebra is isomorphic
to ξ .

Proof

Hochschild [50] ch. 12.   ☐

Now we see how the Lie algebra arises :

A **left-invariant vector field** of a Lie group G is any
w : G $\longrightarrow$ TG satisfying

$$(\forall g,h \epsilon G)\ DL_g \circ w(h)\ =\ w \circ L_g(h)\ =\ w(gh)\ ,$$

that is, fixed under differentials of left translations.   Such
fields form a vector space and in fact a **Lie algebra,** ξ ,   under

the Lie bracket composition

$$[w,v](f) \quad = \quad w(v(f)) - v(w(f))$$

$$\forall w,v \epsilon G, \ \forall C^{\infty} f : G \longrightarrow \mathbb{R} \qquad (cf. \ \S 2.2) .$$

This composition is of course available to all vector fields on G,
i.e. to $T^1 G$, and the inclusion

$$\xi \hookrightarrow T^1 G$$

is a Lie algebra homomorphism.

We can visualise ξ through an isomorphism (in Vec) with the
tangent space to the identity:

$$\xi \equiv T_e G : \gamma \longrightarrow \gamma(e)$$

so we see dim ξ = dim G .

Example 1

In the case $G = G\ell(n;\mathbb{R})$ the tangent space $T_a G$ at any $a \epsilon G$ is
the matrix space

$$T_a G \quad = \quad \{(a,A) | A \ \text{is an} \ n{\times}n \ \text{real matrix}\}$$

so $T_a G \equiv \mathbb{R}^{n^2}$.

The Lie algebra $\xi = \xi\ell(n;\mathbb{R})$ consists essentially of all $n{\times}n$
real matrices A , which determine fields by

$$A : G \longrightarrow TG : a \longrightarrow (a,A)$$

and the Lie bracket is the composite matrix product:

$$[A,B] \quad = \quad AB - BA .$$

We illustrate the left-invariance property for the case n = 1 .
Then $G = \mathbb{R}^*$, $\xi = \mathbb{R}$ and

$$(\forall \gamma \epsilon \mathbb{R}) \ \gamma : \mathbb{R}^* \longrightarrow \mathbb{TR}^* : a \longrightarrow (a,\gamma_a)$$

is left-invariant if

$$(\forall a,g \epsilon \mathbb{R}^*) \ DL_g \gamma(a) = \gamma L_g(a) = \gamma(ga) .$$

Hence at the identity $e = 1 \epsilon \mathbb{R}^*$

$$g\gamma(e) \quad = \quad \gamma(ge) \quad = \quad \gamma(g) ,$$

so we require

$$\gamma(a) \;=\; a\gamma(1) \;=\; a\gamma_1 \;, \quad \text{say} .$$

The adjoint representation Ad of a Lie group G in its Lie algebra ξ is obtained from the derivative of the automorphism

$$ad(g) \;=\; L_g R_{g^{-1}} : G \longrightarrow G : h \longmapsto ghg^{-1} \quad (\forall g \epsilon G)$$

namely,

$$Ad(g) : \xi \longrightarrow \xi : \gamma \longmapsto D(L_g R_{g^{-1}})\gamma = Dad(g)\gamma .$$

It follows that left-invariance is preserved under each map Ad(g). For more theory see Hochschild [50] and Serre [88], and for applications in physics see DeWitt and DeWitt [23].

Example 2

The adjoint representation is trivial for any $g \epsilon G\ell(1;I\!\!R)$; because

$$L_h R_{g^{-1}} : I\!\!R^* \longrightarrow I\!\!R^* : h \longmapsto ghg^{-1} \;=\; h$$

implies $Ad(g) \;=\; I_{I\!\!R}$.

The following is basic to local geometry of Lie groups:

2.5.2   Theorem

Every $\gamma \epsilon \xi$ generates a one-parameter subgroup of G .

Proof (Outline only)

Let $c_\gamma : t \longmapsto \gamma_t$ be the integral curve in G (cf. §2.2), determined for $|t| < \epsilon$ for some real $\epsilon > 0$, with initial conditions

$$\gamma_o \;=\; c_\gamma(0) \;=\; e \epsilon G$$

$$\dot{c}_\gamma(0) \;=\; \gamma(e) \epsilon T_e G \equiv \xi \quad \text{(as a vector space)} .$$

We define

$$\phi_t : G \longrightarrow G : g \longmapsto L_g(\gamma_t), \quad |t| < \epsilon .$$

Since this is well-defined for all $g \epsilon G$ it admits an extension to all $t \epsilon I\!\!R$ and we have a group because where both sides are defined we find

$$L_g(\gamma_{t+s}) \;=\; L_g(\gamma_t) \circ L_g(\gamma_s) .$$

The required subgroup of G is

$$G_\gamma = \{\phi_t(e) \mid t \in \mathbb{R}\} . \qquad \Box$$

Next we obtain the <u>exponential map</u>

$$\exp : \xi \longrightarrow G : \gamma \longmapsto \gamma_1$$

that is, each field is followed along its integral curve to unit distance from the identity (cf. Hochschild [50] p. 79 and Dodson and Poston [28] p. 354.)

Example 3

In the case $G = G\ell(n;\mathbb{R})$ the exponential map coincides with the usual function for matrices γ

$$\exp \gamma = \sum_{k=0}^{\infty} \gamma^k/k! \quad .$$

Something of why this is so can be gleaned from the situation when $n = 1$. Take $\gamma \in \xi\ell(1;\mathbb{R}) = \mathbb{R}$ and seek the integral curve $c_\gamma : (-\varepsilon, \varepsilon) \longrightarrow \mathbb{R}^*$ for which

$$c_\gamma(0) = e = 1$$

$$\dot{c}_\gamma(0) = \gamma(1) = \gamma_o$$

$$\dot{c}_\gamma(t) = \gamma \circ c_\gamma(t) .$$

Evidently, we are seeking a particular solution of the differential equation

$$\frac{d}{dt}c\gamma = \gamma_o \, c_\gamma ,$$

and it is actually

$$c_\gamma : (-\varepsilon, \varepsilon) \longrightarrow \mathbb{R}^* : t \longmapsto e^{\gamma_o t} .$$

Hence the one-parameter subgroup is given by maps of the form

$$\phi_t : \mathbb{R}^* \longrightarrow \mathbb{R}^* : g \longmapsto g \, e^{\gamma_o t}$$

and the exponential map is

$$\exp : \mathbb{R} \longrightarrow \mathbb{R}^* : \gamma \longmapsto e^\gamma .$$

We defined in §1.5 the action of a Lie group G on a manifold; now we can bring in the role of its Lie algebra. Suppose then that G acts on the right of a manifold P

$$P \times G \longrightarrow P : (u,g) \longmapsto R_g(u) .$$

From above we find for each $\gamma \epsilon \xi$ a one-parameter subgroup of G

$$G = \{\gamma_t(e) = \exp t\gamma \,|\, t\epsilon R\} ,$$

acting on the right of P by inclusion in G .   The following properties are readily checked :

(i)   through each $u\epsilon P$ there is a smooth curve

$$\gamma_u : t \longmapsto R_{\gamma_t(e)}(u)$$

with tangent vector $\dot{\gamma}_u(0)\epsilon T_uP$ at $t = 0$.

(ii)   there is a map

$$\Phi : \xi \longrightarrow T^1P : \gamma \longmapsto \dot{\gamma}$$

where T^1P is the Lie algebra of all vector fields on P and

$$\dot{\gamma} : P \longrightarrow TP : u \longmapsto \dot{\gamma}_u \quad \text{(as in (i))}$$

is called the <u>fundamental vector field corresponding to</u> $\gamma\epsilon\xi$.
Then Φ is a Lie algebra homomorphism and if the action of G is free (only e has a fixed point) then $\dot{\gamma}_u$ is not the zero vector at any $u\epsilon P$,   for non-zero $\gamma\epsilon\xi$.

Example 4

Consider $G = O(2;R)$,   the orthogonal subgroup of $G\ell(2;R)$.
Then for any

$$\gamma = \begin{bmatrix} 0 & \theta \\ \theta & 0 \end{bmatrix} \epsilon \xi$$

we have

$$\exp \gamma = \begin{bmatrix} \cos\theta & \sin\theta \\ -\sin\theta & \cos\theta \end{bmatrix} .$$

Example 5

For $G = G\ell(1;R) = R^*$ and any $\gamma\epsilon\xi\ell(1;R) = R$,

$$G_\gamma = \{e^{t\gamma} = \exp t\gamma \mid t \epsilon R\} \subset R^* .$$

Example 6

Using the action of R^* on R^2 given by

$$R^2 {\times} R^* \longrightarrow R^2 : ((x,y),g) \longmapsto (gx,gy)$$

we obtain a curve in R^2 through $u = (x,y)$:

$$\gamma_u : t \longmapsto R_e t\gamma (u) = (xe^{t\gamma}, ye^{t\gamma}) .$$

The tangent vector to this at any t is

$$\dot{\gamma}_u(t) = \left((xe^{t\gamma}, ye^{t\gamma}), (\gamma xe^{t\gamma}, \gamma ye^{t\gamma})\right) \epsilon T_{\gamma_u(t)} R^2 .$$

Evaluating this at $t = 0$ yields the map

$$\gamma : R^2 \longrightarrow TR^2 : u \longmapsto (u, \gamma u)$$

and in this example $\dot{\gamma}_u = \underline{0}$ at $u = \underline{0}$ because all $\gamma \epsilon R$ have a fixed point there.

Let (P,G,M) be a principal fibre bundle. Then for all $\gamma \epsilon \xi$ we note the following properties of the fundamental vector field $\dot{\gamma}$:

(i)   G translates each fibre along itself freely, so the linear map

$$\xi \longrightarrow T_u P : \gamma \longmapsto \dot{\gamma}_u$$

has trivial kernel and is injective for all $u \epsilon P$.

(ii) We can obtain $\dot{\gamma}$ by means of ($\forall u \epsilon P$)

$$\sigma_u : G \longrightarrow P : g \longmapsto R_g(u)$$

for $D\sigma_u : TG \longrightarrow TP$ restricts to give at the identity

$$T_e G \longrightarrow T_u P : \gamma \longmapsto \dot{\gamma}_u$$

where as usual we identify $T_e G$ with ξ .

(iii) Given $\Pi_p(u) = x$

$$D\Pi_p \dot{\gamma}_u = \underline{0} \epsilon T_x M ,$$

i.e. $\dot{\gamma}_u$ is tangent to the fibre containing u .

(iv) As a vector space ξ is isomorphic to the vertical
subspace G_u over any $u \in P$ by the map

$$\xi \longrightarrow G_u : \gamma \longmapsto \dot{\gamma}_u .$$

Hence $\dim G_u = \dim \xi = \dim G$.

Example 7

Take the case of a frame bundle LM with $G = Gl(n,R)$, $\xi = \xi l(n,R)$.
We display the above maps in component form (cf. §2.3 Example 3)
for any $u = (x^i, b^i_j) \in$ LM .

$$\sigma_u : G \longrightarrow LM : (g^i_j) \longmapsto (x^i, g^i_m b^m_j)$$

$$D\sigma_u : T_e G \longrightarrow T_u LM : (\delta^i_j, b^i_j) \longmapsto (x^i, b^i_j, 0, \gamma^i_m b^m_j)$$

$$\xi \longrightarrow T_u LM : (\gamma^i_j) \longmapsto (\gamma^i_m b^m_j)$$

$$(\dot{\gamma}^i_j) : LM \longrightarrow TLM : u \longmapsto (x^i, b^i_j, 0, \gamma^i_m b^m_j) .$$

Note the use of the summation convention for matrix products

$$[g^i_m b^m_j] = [g^i_m] \ [b^m_j] .$$

In particular, for $LS^1 = S^1 \times R^*$ we have $\gamma \in \xi l(1,R) = R$ and
$u = (x,b)$ so we find :

$$\dot{\gamma} : LS^1 \longrightarrow TLS^1 : (x,b) \longmapsto (x,b,0,\gamma b)$$

$$\sigma_u : R^* \longrightarrow LS^1 : g \longmapsto (x,gb)$$

$$D\sigma_u : T_1 R^* \longrightarrow T_u LS^1 : (1,\gamma) \longmapsto (x,b,0,\gamma b) \text{ at } g=1$$

$$T_1 R^* \equiv R : (1,\gamma) \longmapsto \gamma .$$

A Lie algebra ξ is called <u>nilpotent</u> if there is a positive
integer n such that for every n-tuple $(\alpha_1, \alpha_2, \ldots, \alpha_n)$ of
elements from ξ we have

$$\Delta_{\alpha_1} \cdot \Delta_{\alpha_2} \ldots \Delta_{\alpha_n} = 0$$

where

$$\Delta_\alpha : \xi \longrightarrow \xi : \beta \longmapsto (\alpha\beta - \beta\alpha) .$$

2.5.3   Theorem

Let   G   be a connected Lie group with nilpotent Lie algebra ξ .

Then the exponential map is surjective.

If   G   is also simply-connected then the exponential map is
an isomorphism in Man.

Proof

Hochschild [50] p. 136.   □

2.6   *Connection; parallel transport, geodesic, holonomy group*

The early work of Weyl and others on connections (in tangent
bundles) was much stimulated by the growth of interest in general
relativity (cf. Weyl [101] §14 and Notes 9, 10).   More recently
the pure geometric formulation, notably by Ehresmann [32], extended
the notion to general bundles.   Interestingly, these generalizations
have proved very useful in current particle physics under the name
of Yang-Mills field theory (cf. Hermann [45]).   There are three
helpful ways to view a connection in a vector bundle (M,E,Π_E) :

(i)   as a 'covariant derivative' of sections of the bundle   E ,
with respect to tangent vector fields;

(ii)   as a specification of which vector in $\overset{\leftarrow}{\Pi}_E \circ c(t)$ is 'parallel'
to a given vector in $\overset{\leftarrow}{\Pi}_E \circ c(0)$,   with respect to a given curve   c
in   M ;

(iii) as a smooth 'splitting' of the tangent bundle   TE   into
'horizontal' and 'vertical' subspaces over each   u∈E .   A modern
geometric motivation for the definition of a connection is given in
Dodson and Poston [28] Ch. 8, with pictorial representation of the
above three roles.   Our plan for the present section is to define
a connection in a principal fibre bundle then deduce various
properties, including its influence on associated vector bundles.
We shall mainly follow Kobayashi and Nomizu [57] for definitions,
taking some examples from Dodson [26].

A connection ∇ in a principle fibre bundle   (P,G,M)   is an
assignment of a subspace H_u of $T_u P$ for all   u∈P   such that

(i) $T_u P = H_u \oplus G_u$

: $X \longmapsto X_H \oplus X_G$ smoothly on TP, where G_u is the vertical subspace (cf. §2.3) ;

(ii) $DR_g H_u = H_{R_g(u)}$ $\forall g \in G$.

We call H_u the _horizontal subspace_ at $u \in P$. The derivative of Π_P induces an isomorphism

$$H_u \equiv T_{\Pi_P(u)} M \quad (\forall u \in P)$$

and we already have a composite isomorphism (cf. §2.5)

$$G_u \equiv \xi \equiv T_e G \quad (\forall u \in P)$$

Example 1

Consider $P = LS^1$; a _constant_ connection in LS^1 is given by any $\lambda \in \mathbb{R}$ if we put

$$\left(\forall u = (x,b) \in LS^1\right) \; H_{(x,b)} = \{(x,b,p,-\lambda bp) \,|\, p \in \mathbb{R}\} .$$

Then the smooth decomposition of $T_u LS^1$ is $T_u LS^1 = H_u \oplus \mathbb{R}_u^*$:

$$(x,b,p,q) = (x,b,p,-\lambda bp) \oplus (x,b,0,q+\lambda bp) .$$

We check the compatibility with the right action of $g \in \mathbb{R}^*$,

$$R_g : LS^1 \longrightarrow LS^1 : (x,b) \longmapsto (x,bg)$$

$$DR_g : TLS^1 \longrightarrow TLS^1 : (x,b,p,q) \longmapsto (x,bg,p,bg)$$

$$DR_g H_{(x,b)} = H_{(x,bg)} .$$

Also, the differential of the projection Π_L is

$$D\Pi_L : TLS^1 \longrightarrow TS^1 : (x,b,p,q) \longmapsto (x,p)$$

and it gives the isomorphism

$$H_{(s,b)} \equiv T_x S^1 : (x,b,p,-\lambda bp) \longmapsto (x,p) .$$

Of course, $H_{(x,b)}$ only looks horizontal in the standard embedding $LS^1 \hookrightarrow \mathbb{R}^3$ if $\lambda = 0$, which would be the connection giving the usual parallelism structure.

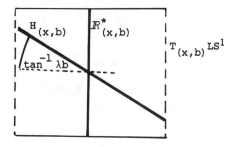

The connection form of a connection ∇ in (P,G,M) is the smooth map

$$\omega : TP \longrightarrow \xi : X_H \oplus X_G \longmapsto \gamma$$

with $X_G(u) = \dot{\gamma}_u$ (cf. §2.5).

Properties of the connection form (cf. Kobayashi and Nomizu [57] p. 64 for proofs.)

(i) $(\forall \gamma \epsilon \xi)$ $\omega(\dot{\gamma}_u) = \gamma$.

(ii) $\omega(X) = \underline{0} \iff X = X_H \oplus \underline{0}$.

(iii) $(\forall g \epsilon G, \forall X \epsilon T^1 P)$

$$\omega \circ DR_g X = Ad(g^{-1})\omega(X), \quad (cf. §2.5) .$$

(iv)   Connections and connection forms determine one another
        uniquely.   □

Example 2

We continue the preceding example:

$$(x,b,p,-\lambda bp) \oplus (x,b,0,q+\lambda bp) = (x,b,p,q) ;$$

since $\dot{\gamma}(x,b) = (x,b,0,\gamma b)$ (cf. §2.5 Ex. 7) ,

$\omega(x,b,p,q) = \gamma = (q+\lambda bp)/b \epsilon \xi\ell(1;\mathbb{R}) = \mathbb{R}$.

We can see the above properties for this example:

(i) $\omega \circ \dot{\gamma}(x,b) = \omega(x,b,0,\gamma b) = \gamma b/b = \gamma$, $(\forall \gamma \epsilon \mathbb{R})$;

(ii) $(q+\lambda bp)/b = 0 \iff q+\lambda bp = 0$.

(iii) Given $X : LS^1 \longrightarrow TLS^1 : (x,b) \longmapsto (x,b,p,q)$, and $g\epsilon R^*$, $Ad(g^{-1})$ is the identity (cf. §2.5) :

$$Ad(g^{-1})\omega(X) \;=\; D(L_{g^{-1}}R_g)\omega(X) \;=\; \omega(X) \ .$$

Next,

$$DR_g : TLS^1 \longrightarrow TLS^1 : (x,b,p,q) \longmapsto (x,bg,p,qg)$$

and

$$\omega(X) : R^* \longrightarrow \mathbb{R}^* : a \longmapsto \big(a,a(q+\lambda bp)/b\big)$$

so, as required,

$$\omega\circ DR_g(X) : R^* \longrightarrow \mathbb{R}^*$$

satisfies property (iii) :

$$\omega\circ DR_g(x,b,p,q) \;=\; \omega(x,bg,p,qg)$$

$$=\; (qg+\lambda bgp)/bg \;=\; (q+\lambda bp)/b.$$

(iv) Evidently ω and λ determine one another uniquely.   Next we come to an important construction process:

2.6.1   Theorem

Given a connection ∇ in a principal fibre bundle (P,G,M) , every vector field $w : M \longrightarrow TM$ has a unique <u>horizontal lift</u>

$$w^\uparrow : P \longrightarrow TP : u \longmapsto w^\uparrow(u)$$

with the following properties.

(i) $D\Pi_p w^\uparrow(u) \;=\; w\circ\Pi_p(u)$ and $w^\uparrow(u)\epsilon H_u$.

(ii) $w^\uparrow$ is invariant by R_g $(\forall g\epsilon G)$, and every horizontal vector field on P invariant by G is the horizontal lift of some vector field on M .

(iii) $w^\uparrow + v^\uparrow \;=\; (w+v)^\uparrow$.

(iv) If $f\epsilon T^0 M$ we put $f^\uparrow = f\circ\Pi_p$

then for any vector field $w : M \longrightarrow TM$

$$(f.w)^\uparrow \;=\; f^\uparrow.w^\uparrow : P \longrightarrow TP \ .$$

(v) $[v^\uparrow, w^\uparrow]_H = [v,w]^\uparrow$ (H denotes horizontal component).

Proof

Kobayashi and Nomizu [57] p. 65.   □

Corollary

Every piecewise c^1- curve c : $[0,1) \longrightarrow M$ has a unique
horizontal lift curve

$$c^\uparrow : [0,1) \longrightarrow P, \text{ through } u_0 \epsilon \overleftarrow{\Pi}_P \circ c(0) ,$$

with $\Pi_P \circ c^\uparrow = c$, $c^\uparrow(0) = u_0$;
and the tangent vector field of $c^\uparrow$ is the horizontal lift of $\dot{c}$,
that is $\dot{c}^\uparrow$.

Moreover, the map

$$\tau_t : \overleftarrow{\Pi}_P \circ c(0) \longrightarrow \overleftarrow{\Pi}_P \circ c(t) : u_0 \longmapsto c^\uparrow(t) ,$$

called parallel transport along c , commutes with the action of
G on P and is a diffeomorphism.

Proof

Kobayashi and Nomizu [57] p. 68-70,   cf. also Dodson and Poston
[28] Ch. 8 for more discussion .   □

     We shall look at examples of these horizontal lifts shortly
but first it is convenient to interpret the connection in (P,G,M)
as a derivation in an associated vector bundle (cf. §2.1 and §2.3).
For this purpose we must extend the notion of parallel transport
in a principal fibre bundle to its associated vector bundles.
(Note that it is not difficult to make any vector bundle an
associated bundle of some principal bundle.)
     Let ∇ be a connection in (P,G,M) and let (M,E,Π_E) be a
vector bundle with fibre V associated to (P,G,M).   We effect a
splitting of each tangent space $T_a E$ for $a \epsilon E$ into a direct sum

$$T_a E = H_a \oplus G_a ,$$

with $G_a = \{X \epsilon T_a E \mid D\Pi_E X = 0 \epsilon T_{\Pi_E(a)} M\}$

as a vertical subspace, and H_a the horizontal subspace defined
as follows (cf. §2.3).   Denote the canonical factorisation by G

of P×V by:

$$f_G : P×V \longrightarrow (P×V)/G = E$$

and choose any $(u,y) \epsilon f_G^{\leftarrow}(a)$, that is $[(u,y)]_G = a$.
Fix y and introduce the map

$$f : P \longrightarrow E : v \longrightarrow f_G(v,y) .$$

The derivative of this map at $u \epsilon P$ takes us to $T_a E$

$$D_u f : T_u P \longrightarrow T_a E$$

and we define the horizontal subspace of $T_a E$ by

$$H_a = D_u f(H_u) .$$

This construction makes H_a independent of the choice of (u,y)
and does indeed yield the required direct sum with G_a ; it
varies smoothly with a because G_a varies smoothly, and so
does H_u with u .   Now we can obtain unique horizontal lifts
from M to E in much the same way as we did from M to P .
Hence, we obtain also parallel transport between fibres of E
along a given curve c in M :

$$\tilde{\tau}_t : \Pi_E^{\leftarrow} \circ c(0) \longrightarrow \Pi_E^{\leftarrow} \circ c(t)$$

$$: a_o \longrightarrow f_G(c^{\uparrow}(t),y)$$

where y is any element of V with $a_o = [(u_o,y)]_G$, and $c^{\uparrow}$ is
the horizontal lift of c to P .   As before this map is a
diffeomorphism and since here the domain and image are vector
spaces it is an <u>isomorphism</u>.

    We say that a section v of E defined on an open set U
of M is <u>parallel along a given curve</u> $c : [0,1] \longrightarrow U$, if
we have

$$\tilde{\tau}_t \circ v \circ c(0) = v \circ c(t) ;$$

if v is parallel along <u>all</u> curves in U we say that v is a
<u>parallel section</u>.

    Let (P,G,M) be a principal fibre bundle with connection ∇
and let (M,E,Π_E) be a vector bundle with fibre V associated to
(P,G,M) .   Given any C^1- curve c : $[0,1) \longrightarrow$ M and a section v

of E defined along c , that is

$$\Pi_E \circ v \circ c(t) \ = \ c(t) \qquad \forall \ t\epsilon[0,1) \ ,$$

then the <u>covariant derivative</u> of v along c is

$$\nabla_{\dot{c}(t)} v \ = \ \lim_{h \to 0} \frac{1}{h}\left(\overset{\leftarrow}{\tau_h} v \circ c(t+h) - v \circ c(t)\right)$$

where τ_h is the parallel transport isomorphism

$$\tau_h : \overset{\leftarrow}{\Pi_E} \circ c(t) \ \longrightarrow \ \overset{\leftarrow}{\Pi_E} \circ c(t+h) \ .$$

Hence, $\forall t\epsilon[0,1)$, $\nabla_{\dot{c}(t)} v \epsilon \overset{\leftarrow}{\Pi_E} \circ c(t)$ so defining a section of E
along c .

Properties of the covariant derivative

(i)   v is parallel along c $\Longleftrightarrow$ $\nabla_{\dot{c}(t)} v = \underline{0} \ \forall t$.

(ii) $\nabla_{\dot{c}(t)} (v+w) \ = \ \nabla_{\dot{c}(t)} v + \nabla_{\dot{c}(t)} w$.

(iii)   If f is a real function defined along c then

$$\nabla_{\dot{c}(t)} (f \, v) \ = \ f \circ c(t) . \ \nabla_{\dot{c}(t)} v + \dot{c}(t)(f).v \circ c(t) \ .$$

(iv)   We can extend the application to E_U, sections of E
defined on some open neighbourhood U of xϵM , hence obtaining
a map

$$\nabla_x : T_x M \times E_U \ \longrightarrow \ E_U$$

$$: (r,v) \ \longmapsto \ \nabla_{\dot{c}(0)} v \ = \ \nabla_r v, \text{ say },$$

where c is any curve in M with $\dot{c}(0) = r$.

(In particular, we note the important case when E_U is one of the
spaces of tensor fields $T^k_h M$, Example 3 below.)
Using this notation we further obtain:

(v) $\nabla_{r+s} v \ = \ \nabla_r v + \nabla_s v$.

(vi) $\nabla_r (v+w) \ = \ \nabla_r v + \nabla_r w$.

(vii) $\nabla_{\lambda r} v \ = \ \lambda \nabla_r v \qquad (\forall \lambda \epsilon I\!R)$.

(viii) $\nabla_r f v \ = \ f(x) \nabla_r v + r(f).v(x)$, $(\forall f : U \to I\!R)$.

(ix)   If r and s are vector <u>fields</u> on M , i.e. r,s∈TM ,
and v,w are sections of E on M then properties (v) - (viii)
remain valid but the expressions are now <u>sections</u> of E on M .

<u>Proof</u>

Kobayashi and Nomizu [57] Ch. 3, Dodson and Poston [28] Ch. 8.   ☐

<u>Example 3</u>

In the particular case of a connection ∇ in a frame bundle the
induced covariant derivation acts on tensor fields on M .   We
display the action in local co-ordinates for tangent vector fields.
Suppose that (U,φ) is a chart on M , then we have a local
field of frames

$$(\partial_i) : U \longrightarrow LM : x \longmapsto (\partial_i)_x$$

with respect to which any field v : M ⟶ TM has a local
expression

$$v(x) = v^i(x)\partial_i(x) \qquad \forall x \in U .$$

We can safely omit reference to the point x and write
$v = v^i \partial_i \in T^1 U$.   Now, the covariant derivation is linear and so it
is uniquely determined locally in U by its action on the basis
fields (∂_i) :

$$\nabla_{\partial_i} \partial_j = \Gamma^k_{ij} \partial_k \qquad \forall\, i,j .$$

The arrays of real functions Γ^k_{ij} are the <u>Christoffel symbols</u> ,
or <u>connection components for</u> ∇ with respect to the chart (U,φ) .
        We use these Γ^k_{ij} to display how ∇ splits each $T_u LM$.
Firstly we note that u∈LM admits a local expression :
$u = (x^i, b^i_j) \in R^n \times R^{n^2}$ where $\Pi_L(u) = x \in U$ and $\phi(x) = (x^i) \in R^n$ and
the frame determined by u for $T_x M$ is

$$b^i_1 \partial_i, \ b^i_2 \partial_i, \ \ldots, \ b^i_n \partial_i .$$

Now, given $Y = (x^i, b^i_j, X^i, B^i_j) \in T_u LM$

we have (cf. Examples 1,2 for the case n = 1):

$$Y = Y_H \oplus Y_G = (x^i, b^i_j, X^i, -b^j_{\cdot} \Gamma^i_{k\ell} X^\ell) \oplus (x^i, b^i_j, 0, B^i_j + b^k_j \Gamma^i_{k\ell} X^\ell) .$$

Also, in local co-ordinates, the connection form ω sends :

Y to that matrix $[\gamma^i_j] = \gamma \epsilon \xi \ell(n;\mathbb{R})$ such that:

$$\dot{Y}_u = Y_G, \text{ that is } [\gamma^i_j] = [B^i_j + b^k_j \Gamma^i_{k\ell} x^\ell][b^i_j]^{-1} .$$

Omitting the indices from these matrices we write

$$\gamma = (B + b\Gamma X) b^{-1} .$$

Now we display the property

$$(\forall g \epsilon G) \, \omega \circ DR_g Y = Ad(g^{-1}) \gamma .$$

We have $DR_g(Y)$, for $g = [g^i_j] \epsilon G\ell(n;\mathbb{R})$, given by

$$(x^i, g^i_m b^m_j, x^i, -g^k_m b^m_j \Gamma^i_{k\ell} x^\ell)$$

$$\ominus \ (x^i, g^i_m b^m_j, 0, g^i_m B^m_j + g^k_m b^m_j \Gamma^i_{k\ell} x^\ell) .$$

Then

$$\omega \circ DR_g Y = \eta = [\eta^i_j] \text{ such that}$$

$$(\eta^i_k g^k_m b^m_j) = (g^i_m B^m_j + g^k_m b^m_j \Gamma^i_{k\ell} x^\ell) .$$

Omitting indices from the matrices and solving for η :

$$\eta = (gB + gb\Gamma X)(gb)^{-1}$$

$$= g(B + b\Gamma X) b^{-1} g^{-1}$$

$$= R_{g^{-1}} \circ L_g(\gamma) = Ad(g^{-1}) \gamma .$$

Next we find the local differential equation for the horizontal lift $c^\uparrow$ of a curve

$$c : [0,1] \longrightarrow U \text{ which locally appears as}$$

$$\phi \circ c : [0,1] \longrightarrow \mathbb{R}^n : t \longmapsto (c^i(t)) ,$$

where (U, ϕ) is some chart on M . We have

$$c^\uparrow : [0,1] \longrightarrow LU , \quad \Pi_L \circ c^\uparrow = c .$$

Locally,

$$\dot{c} = \dot{c}^i \partial_i \text{ with } \dot{c}^i = \frac{dc^i}{dt} ,$$

and since each $c^\uparrow(t)$ is a frame at $c(t)$ it can also be expressed via the frame $(\partial_i)_{c(t)}$:

$$\vec{c}(t) = (b^i_j \partial_i)_{c(t)} \; .$$

Now, for each $j = 1,2,\ldots,n$: $b^i_j \partial_i \epsilon T_{c(t)} M$, we want it to be parallel along c and this means

$$\nabla_c (b^i_j \partial_i) = \underline{0} \; , \quad j = 1,2,\ldots,n \; .$$

From the properties of ∇ this equation becomes

$$\nabla_{\overset{\cdot\rightarrow}{c}\partial_i}{}^{(b^i_j\partial_i)} = \left(\frac{d}{dt}\, b^i_j + b^k_j \overset{\cdot}{c}{}^\ell \Gamma^i_{k\ell} \right) \partial_i = \underline{0} \; ,$$

so the differential equations to be solved are

$$\frac{d}{dt}\, b^i_j + b^k_j \overset{\cdot}{c}{}^\ell \Gamma^i_{k\ell} = 0 \; , \quad i,j = 1,2,\ldots,n.$$

These do have a unique solution $b^i_j(t)$ satisfying given initial conditions

$$[b^i_j(0)] = [\beta^i_j]$$

by the classical Cauchy theorem (note that the specification of $\vec{c}(0)$ also fixes $\overset{\cdot\rightarrow}{c}(0)$ via the equation). Then we have the parallel transport map along c :

$$\tau_t : \Pi^{\leftarrow}_L \circ c(0) \longrightarrow \Pi^{\leftarrow}_L \circ c(t)$$

$$: (\beta^i_j \partial_i) \longmapsto (b^i_j(t) \partial_i) \; .$$

Example 4

Further to Examples 1, 2, consider the case of the constant connection $\lambda = \Gamma^1_{11} \epsilon R$ in the frame bundle LS^1 . The curve (with fixed $\alpha \epsilon R$)

$$c : [0,1) \longrightarrow S^1 : t \longmapsto \alpha t$$

where we consider S^1 as R (mod 1) ,
has horizontal lift

$$\vec{c} : [0,1) \longrightarrow LS^1 : t \longmapsto (\alpha t, b(t))$$

with

$$\frac{db}{dt} + b\lambda\alpha = 0 \ ,$$

and $b(0) = b_0$ say, for some $b : [0,1) \longrightarrow R^*$. Hence
$b(t) = b_0 e^{-\lambda\alpha t}$ and parallel transport here is

$$\tau_t : \Pi_L^\leftarrow \circ c(0) \longrightarrow \Pi_L^\leftarrow \circ c(t) : (0,b_0) \longmapsto (\alpha t, b_0 e^{-\lambda\alpha t}) \ .$$

Since $b(t)$ cannot be zero, $c^\uparrow(t)$ spirals round one component
of LS^1 (if $\alpha \neq 0$), while c proceeds indefinitely round S^1 .

Example 5

A connection ∇ in a frame bundle LM determines the geodesic
curves, their tangent vectors are parallel along the curves:

$$c : [0,1) \longrightarrow M \quad \text{with} \quad \nabla_{\dot c} \dot c = \underline{0} \ .$$

In local coordinates their differential equations are

$$\frac{d\dot c^i}{dt} + \dot c^j \dot c^k \Gamma_{jk}^{\ i} = 0 \ , \quad i = 1,2,\ldots,n \ .$$

We return now to a connection ∇ in a principal fibre bundle
(P,G,M) to define the holonomy groups of ∇ and here 'curve'
will mean a piecewise smooth curve.

The loop space $C(x)$ at any $x\epsilon M$ is the set of all closed
curves in M starting and ending at x ; there is a natural
product on $C(x)$. The holonomy group of ∇ with reference point
$x\epsilon M$ is

$$H(x) = \{\tau_c | c\epsilon C(x)\}$$

where τ_c is the parallel transport automorphism around c ,

$$\tau_c : \Pi_P^\leftarrow(x) \longrightarrow \Pi_P^\leftarrow(x) \ .$$

(Note that τ_c commutes with the action of G on the fibre
$\Pi_P^\leftarrow(x)$.) We can realise $H(x)$ as a subgroup of G , namely

$$H(u) = \{g\epsilon G | R_g(u) = \tau_c(u) \ , \ \tau_c \epsilon H(x)\} \ ,$$

the holonomy group of ∇ with reference point $u\epsilon P$.
If we write for $u,v\epsilon P$,

$$u \sim v \iff (\exists \text{ horizontal smooth curve in } P \text{ from } u \text{ to } v)$$

then $\sim$ is an equivalence relation, partitioning P into disjoint non-empty sets and we have

$$H(u) = \{g \in G \mid u \sim R_g(u)\}$$

Properties of the holonomy group $H(u)$

(i) $(\forall g \in G)\ H(R_g(u)) = ad(g^{-1})H(u)$.

(ii) $u \sim v \Longrightarrow H(u) = H(v)$.

(iii) If M is connected and paracompact then $H(u)$ is a Lie
subgroup of G , and $P(u) = \{v \in P \mid u \sim v\}$ gives a principle fibre
bundle over M with structure group $H(u)$.   The bundle P is
decomposed into the disjoint union of these holonomy bundles $P(u)$,
$u \in P$.

(iv) The restricted holonomy group of ∇ with reference point u
is the identity component of $H(u)$:

$$H^o(u) = \{g \in H(u) \mid R_g(u) = \tau_c(u),\ c \text{ homotopic to zero in } C(x)\}.$$

(v) $u \sim v \Longrightarrow H^o(u) = H^o(v)$.

(vi) If M is connected and paracompact then $H^o(u)$ is a
connected Lie (normal) subgroup of G and $H(u)/H^o$ is countable.

Proof

Kobayashi and Nomizu [57] p. 71-75, 84-85.   □

Remark

The action of G on P preserves horizontality and so $(\forall g \in G)$
we get a bundle isomorphism:

$$R_g : P(u) \longrightarrow P(R_g(u))$$

$$ad(g^{-1}) : H(u) \longrightarrow H(R_g(u)) .$$

Since, $(\forall u, v \in P)(\exists g \in G) : P(v) = P(R_g(u))$,
the holonomy bundles $P(u)$, $u \in P$ are all isomorphic.

Example 6

Consider $I\!R$ with the constant connection λ . Horizontal curves
are essentially the same as for LS^1 (cf. Example 4). Given any

$$c : [0,1] \longrightarrow R : t \longmapsto x + \alpha t$$

beginning at x , with α fixed in R , we find

$$c^\dagger : [0,1] \longrightarrow I\!R : t \longmapsto (x + \alpha t, be^{\alpha t})$$

for the unique horizontal lift through $(x,b) \epsilon I\!R$.
Now, closed curves in R , beginning and ending at $x \epsilon R$ are
essentially of the form

$$c : \begin{cases} t \longmapsto x + \alpha t & t \in [0,\tfrac{1}{2}) \\ \\ t \longmapsto x + \alpha(1-t) & t \in [\tfrac{1}{2},1] \, . \end{cases}$$

Therefore parallel transport τ_c for $c\epsilon C(x)$ is the identity on
$\Pi_L^\leftrightarrow(x)$. The action of R^* on $I\!R$ is free and so, $\forall (x,b)\epsilon I\!R$,

$$H(x,b) = \{g\epsilon R^* | R_g(u) = u\} = \{1\} \, .$$

Hence the holonomy groups are all trivial and the holonomy bundle
through (x,b) is
simply the exponential
curve $y = be^{\alpha t}$, $t\epsilon R$,
(in $L^+\!R$ if $b>0$, in
$L^-\!R$ if $b<0$) .
We see that such curves
certainly partition $I\!R$,
since they do not
intersect but every
point of $I\!R$ lies on
one of them.

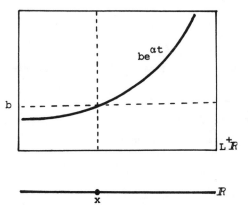

Example 7

Here we see that the holonomy group for LS^1 with constant
connection λ is not trivial. For any $x\epsilon S^1 = R(\text{mod } 1)$ the
loop space $C(x)$ still contains trivial curves, as for R , but
now we have also non-trivial curves

$$c_k : [0,k] \longrightarrow S^1 : t \longmapsto (x+t)(\text{mod } 1) \quad \text{for any}$$

$k = \pm 1, \pm 2, \ldots$.

It turns out that

$$H(x) = \{\tau_{c_k} | k \in Z\}$$

$$\tau_{c_k} : (x,b) \longmapsto (x,be^{-\lambda k}),$$

so, as a subgroup of R^* we
have

$$H(x,b) = \{e^{-\lambda k} \in R^* | k \in Z\} .$$

The identity component is
trivial, $H^o(x,b) = \{1\}$, so
property (vi) is trivially
true. The horizontality
relation on LS^1 is

$$(x,b) \sim (x,be^{-\lambda k}) \quad \text{if} \quad k \in Z .$$

We can check property (i) above by $(\forall g \in R^*)$

$$H(R_g(x,b)) = H(x,bg) = H(x,b)$$

$$\text{ad}(g^{-1})H(x,b) = H(x,b), \quad (\text{cf. } \S2.5 \text{ Example 2}) .$$

Example 8

The connection in IR^2 with:

$$\Gamma^1_{11}(x,y) = y, \ \Gamma^2_{22} = -x, \quad \text{otherwise} \quad \Gamma^k_{ij} = 0, \text{ for all} \quad (x,y) \in R^2$$

is discussed in Dodson [26] p.440. At $(0,0) \in R^2$ the restricted
holonomy group is R^+ .

Suppose that (P',G',M) is a reduced bundle (cf. $\S2.3$) of
the principle fibre bundle (P,G,M) and that ∇ is a connection
in (P,G,M) . Then the connection ∇ is <u>reducible</u> to a
connection ∇' in (P',G',M) if the inclusion morphism

$$(P',G',M) \hookrightarrow (P,G,M)$$

preserves the connection structure (horizontal subspaces, connection
form, holonomy groups etc., cf. Kobayashi and Nomizu [57] p. 81).

2.6.2   Theorem

Given a connection ∇ in a principal fibre bundle (P,G,M) over
a connected paracompact manifold M then each holonomy bundle
$P(u)$ is a reduced bundle with structure group $H(u)$ and ∇ is
reducible to a connection in $P(u)$.

Proof

Kobayashi and Nomizu [57] p. 84.   □

Example 9

We continue Example 7.   For any $(x,b) \in L^+S^1$ we have

$$Q = L^+S^1 (x,b) = \{ (x \overset{+}{+} t, be^{-\lambda t}) \mid t \in R \}$$

here $\overset{+}{+}$ is addition mod 1 .   The projection map is

$$\Pi_Q : (x \overset{\hat{+}}{} t, be^{-\lambda t}) \longrightarrow x \overset{\hat{+}}{} t$$

and the fibre over $x \in S^1$ is

$$\Pi_Q^+(x) = \{ (x, be^{-\lambda k}) \mid k \in Z \} .$$

The reduced bundle is then (Q,Z,S^1) where the right action of Z
is

$$Q \times Z \longrightarrow Q : ((y,d),k) \underset{R_k}{\longrightarrow} (y, de^{-\lambda k}) ,$$

evidently a free action since

$$de^{-\lambda k} = d \implies k = 0 .$$

Following §2.3 we need only provide an open cover $\{U_\alpha \mid \alpha \in A\}$ of S^1
and a family $\{\psi_{\alpha\beta} \mid \alpha, \beta \in A\}$ of transition functions to show that Q
is a principle fibre bundle to which L^+S^1 is reducible.   Take the
cover : $U_\alpha = S^1$ $(\forall \alpha \in Z)$

and the diffeomorphisms

$$\psi_\alpha : \Pi_Q^+ U_\alpha \longrightarrow U_\alpha \times Z : (r, be^{-\lambda m}) \longrightarrow (r, m+\alpha) .$$

Then, putting $(\forall \alpha \in Z)$

$$\phi_\alpha : \Pi_Q^+ U_\alpha \longrightarrow Z : (r, be^{-\lambda m}) \longrightarrow m+\alpha$$

we have $(\forall k \in Z)$ as required

$$\phi_\alpha \circ R_k = R_k \circ \phi_\alpha .$$

Suppose now that

$$(r,be^{-\lambda m}) \in \Pi_Q^+(U_\alpha \cap U_\beta) \ = \ Q \ ,$$

then we find

$$\left(\phi_\beta \circ R_k(r,be^{-\lambda m})\right)\left(\phi_\alpha \circ R_k(r,be^{-\lambda m})\right)^{-1}$$

$$= \ (m+k+\beta) \ - \ (m+k+\alpha) \ = \ \beta-\alpha$$

$$= \ \left(\phi_\beta(r,be^{-\lambda m})\right)\left(\phi_\alpha(r,be^{-\lambda m})\right)^{-1} \ = \ (m+\beta)-(m+\alpha) \ = \ \beta-\alpha \ .$$

Hence suitable transition functions are (cf. §2.3)

$$\psi_{\beta\alpha} \ : \ U_\alpha \cap U_\beta \ \longrightarrow \ Z \ : \ r \ \longmapsto \ \beta-\alpha \ ;$$

that is,

$$\psi_{\beta\alpha} \circ \Pi_Q(r,be^{-\lambda m}) \ = \ \left(\phi_\beta(r,be^{-\lambda m})\right)\left(\phi_\alpha(r,be^{-\lambda m})\right)^{-1} \ .$$

This is well-defined because given $r \in dom(\psi_{\beta\alpha})$ it does not matter to which $(r,be^{-\lambda m})$ we lift it in Q before we apply $\psi_{\beta\alpha} \circ \Pi_Q$.

Now Q is a one-dimensional manifold, an exponential spiral wrapped around the cylinder. Hence at any $(x,b) \ = \ u \in Q$ the vertical subspace Z_u is trivial (cf. §2.3). Hence we have only one choice for a connection ∇' in Q , it has horizontal subspace

$$H_u \ = \ T_u Q \ \ (\forall u \in Q) \ .$$

Then as required, since $(\forall k \in Z)$, R_k is linear, $DR_k H_u = H_{R_k(u)}$.

It follows that ∇ (i.e. λ in LS^1) is reducible to ∇' in Q . Evidently Q was constructed from (and only from) the horizontal curves through $(x,b) \in L^+ S^1$ so the inclusion $Q \hookrightarrow LS^1$ preserves the horizontal structure. The connection form of ∇' is trivial, because the Lie algebra of Z is trivial and the holonomy groups of ∇' are again copies of Z .

2.6.3   Theorem

Given a principal fibre bundle (P,G,M) a closed subgroup G' of G and the associated bundle $E = P/G'$ with fibre G/G' , suppose that $\sigma : M \longrightarrow E$ is a section. Then there is a reduced bundle

(P',G',M) corresponding to σ and a connection ∇ in P is reducible to a connection ∇' in P' if and only if σ is parallel with respect to ∇ .

Proof

Kobayashi and Nomizu [57] p.88 . The correspondence : $\sigma \longrightarrow P'$ is one-to-one and derives from the proof of Theorem 2 in §2.3.   □

2.6.4   Theorem

Suppose that M is parallelizable by the section

$$p : M \longrightarrow LM : x \longmapsto (p_i)_x$$

then p determines a connection ∇^p in LM such that

$$\nabla^p_{p_i}(p_j) = \underline{0} \qquad \forall\ i,j = 1,2,\ldots,n.$$

If

$$h : M \longrightarrow G\ell(n;\mathbb{R}) : x \longmapsto [h^i_j]_x$$

is a smooth map then

$$q : M \longrightarrow LM : x \longmapsto (q_i)_x\ ,\ q_i = h^k_i p_k\ ,$$

is another parallelization and

$$\nabla^q = \nabla^p \quad <=> \quad h \text{ is constant on connected components of } M.$$

Proof

The Christoffel symbols, with respect to the frame $(p_k)_x$, for ∇^p all vanish by hypothesis:

$$\nabla^p_{p_i}(p_j) \ =\ \Gamma^k_{ij}p_k \ =\ \underline{0} \ => \ \Gamma^k_{ij} \ =\ 0\ \forall\ i,j,k\ .$$

So from Example 3 we find that locally the splitting of T_uLM ,
$\forall u = \left(x, (b^i_j p_i)_x\right) \in LM$, is given by

$$T_uLM \ =\ H_u \oplus G_u$$

$$: \left(x, (b^i_j p_i)_x, X, B\right) \ =\ \left(x, (b^i_j p_i)_x, X, 0\right) \oplus \left(x, (b^i_j p_i)_x, 0, B\right)\ .$$

So we choose

$$H_{p(x)} \ =\ D_x p(T_x M)$$

and for any $u \in LM$ with $u = R_h p(x)$

$$H_u = D_{p(x)} R_h (H_{p(x)}) \; .$$

We know that $LM = M \times G\ell(n;\mathbb{R})$ by the existence of p, now we see that the connection ∇^p makes the horizontal subspaces look horizontal in this product bundle by

$$LM \longrightarrow M \times G : \left(x, (b_j^i p_i)_x \right) \longrightarrow \left(x, \delta_j^i \right) \; ,$$

i.e. we locate the identity in G at the frame determined by the parallelization at each point .

Evidently the given q is another parallelization and

$$\nabla^p_{p_i} q_j = \nabla^p_{p_i} h_j^k p_k = p_i (h_j^k) p_k + h_j^k \nabla^p_{p_i} p_k$$

which equals $\underline{0}$ if and only if h_j^k is constant on each connected component of M . $\square$

Corollary 1

The connection ∇^p need not be symmetric (proved in Theorem 8 §2.8) . $\square$

Corollary 2

The geodesics of ∇^p are the integral curves of constant linear combinations like

$$X : M \longrightarrow TM : x \longrightarrow a^i p_i \; .   \square$$

2.6.5   Theorem

Suppose that M is a connected, paracompact n-manifold (for some n>1) with a principal fibre bundle (P,G,M) then:

(i)   There always exists a connection in P, having holonomy bundles coinciding with P .

(ii)   If the given bundle is reducible to (P',G',M) then there is a connection in P having each holonomy bundle coinciding with P' .

(iii) If (P',G',M) is a subbundle of (P,G,M) then a connection ∇ in P is reducible to a connection in P' if for all $u \in P'$ the horizontal subspace of $T_u P$ is tangent to P' .

(iv) Any connected Lie group H can be realized as the holonomy
group of some connection in Q = M×H .

Proof

(i)   A trivial bundle U×G is constructed over a neighbourhood
U of some point in M and provided with a connection having
holonomy group the identity component of G . This is
extendible to P by means of a partition of unity on M , available
because M is paracompact.   Details are given in Kobayashi and
Nomizu [57] p.90–91 (cf. also Brickell and Clark [9] p. 154), the
construction also proves (iv).

(ii) Follows from (i) because the reduced bundle is also a
principal fibre bundle.

(iii) Kobayashi and Nomizu [57] p. 85 : define ∇' in P' by
$H'_u = H_u$, where H_u is the horizontal subspace determined in
$T_u P$ by ∇ and $u \epsilon P' \subseteq P$.   □

Corollary

The structure group $G\ell(n;\mathbb{R})$ of LM can be reduced to a subgroup
G' if and only if there exists a connection in LM having G'
as its holonomy group.   (Cf. Theorem 9 in Ch. V §1.)

Proof

By (ii), if LM is reducible then there is such a connection.
Conversely, LM is always reducible to the holonomy bundle of a
given connection.   This result is originally due to Hano and
Ozeki [42].   □

From the construction of the holonomy bundle one might expect that
it is the 'smallest' reduced subbundle to which a connection is
reducible.   This is indeed the case as we see in the next result.

2.6.6   Theorem

Suppose that M is a connected, paracompact n-manifold (for
some n>1) and ∇ is a connection in a principal fibre bundle
(P,G,M) . If ∇ is reducible to a connection ∇' in a reduced

subbundle (P',G',M) then for any $q \epsilon P'$ the holonomy bundle $P(q)$ is a reduced subbundle of P' and ∇' is reducible to ∇^H , the reduction of ∇ to $P(q)$.

Proof   (Due to M.J. Slupinski)

We need to show that :

(i)   there is a smooth inclusion $h : P(q) \hookrightarrow P'$;

(ii)   there is a Lie group map $i : H(q) \hookrightarrow G'$;

(iii) P' is reducible to $P(q)$;

(iv) ∇' is reducible to ∇^H .

(i)   Suppose $p \epsilon P(q)$. Then by definition there is a ∇-horizontal curve $\bar{c}$ in P from q to p . This curve projects to give $c = \Pi_p \circ \bar{c}$ in M and c has a unique ∇'-horizontal lift c' in P' . By hypothesis there is a smooth inclusion $j : P' \hookrightarrow P$ which also induces the reduction of ∇ to ∇' . It follows that $j \circ c' = c^\uparrow$, the unqiue ∇-horizontal lift of c to P and hence $j \circ c' = \bar{c}$. Therefore $p \epsilon P'$ and so $P(q) \subseteq P'$.

(ii) We know that $P(q)$ is a reduced subbundle of P with structure group $H(q)$ and ∇ is reducible to ∇^H , by Theorem 2.   We have Lie group inclusions

but so far we only have a group inclusion

$$i : H(q) \hookrightarrow G'$$

since $P(q) \subseteq P'$. However, by the Proposition at the end of §2.2 this map i is smooth because the connected components of G' are

integral manifolds of the distribution on G defined by the Lie
algebra of G' , and G' is second countable as a Lie subgroup.

(iii) For this we use Lemma 1 in Kobayashi and Nomizu [57] p. 84.

(iv) We know that ∇ reduces to ∇' in P' and to ∇^H in P(q),
and P(q) is a subbundle of P' .   Now, if a curve c is
∇'-horizontal in P' through $p\epsilon P(q) \subseteq P'$ then it is ∇-horizontal
in P and hence it lies in P(q) .   So, by Lemma 2 in Kobayashi
and Nomizu [57] p. 84, ∇' reduces to some connection in P(q)
and by uniqueness of reductions it must be ∇^H . $\Box$

2.7   *Differential forms;   curvature, torsion*

Certain features of a connection are conveniently expressed in
the language of differential forms, that is in terms of purely
antisymmetric tensor fields.   There is a natural derivation on
such fields that preserves anti-symmetry, the exterior derivative.
Introductory treatments of the so-called exterior (or Grassmann)
algebra and calculus can be found in Spivak [93], Bishop and
Goldberg [7], Singer and Thorpe [90], Cartan [11];   further
developments and applications are given in Flanders [33] and
Kobayashi and Nomizu [57].   Here we shall give about the minimum
for our needs, partly because of the excellently detailed treatments
mentioned above and partly because it is in the nature of differential
forms to use paper extravagantly.

Exterior algebra

Denote by T_r the tensor space of r-linear real valued maps on a
vector space V and by ΛT_r denote the vector subspace of purely
anti-symmetric members of T_r .   Then we have the following

(i)   dim $T_r = n^r$ and dim $\Lambda T_r = \binom{n}{r}$ if dim V = n .

(ii)   There is an alternating operator for each $r = 0,1,\ldots$

$$A_r : T_r \longrightarrow \Lambda T_r : W \longmapsto W_a$$

where for all $v_1, v_2, \ldots, v_r \epsilon V$

$$W_a(v_1, v_2, \ldots, v_r) = \frac{1}{r!} \sum_{\sigma \in S_r} \text{sgn } \sigma \; W(v_{\sigma(1)}, v_{\sigma(2)}, \ldots, v_{\sigma(r)}) \; .$$

(Here S_r is the symmetric group of permutations on the set $\{1,2,\ldots,r\}$ and sgn σ is the parity homomorphism (cf. Macdonald [65] p. 244) to the multiplicative group on $\{-1,1\}$. Recall that the determinant function is defined similarly - and it is no accident (cf. [28] p. 45 et seq.):

$$\det [a_{ij}] = \sum_{\sigma \in S_r} \text{sgn } \sigma \; a_{1\sigma(1)} \cdot a_{2\sigma(2)} \cdot \ldots \cdot a_{r\sigma(r)} \; .)$$

(iii) There is an associative, anticommutative, distributive <u>exterior product</u>

(iv) If (b^i) is a basis for $\Lambda T_1 = T_1$ then a basis for ΛT_r is

$$\{b^{i_1} \wedge b^{i_2} \wedge \ldots \wedge b^{i_r} | i_1 < i_2 \ldots < i_r\} \; .$$

(v) From the properties of binomial coefficients we observe that for each $m = 0,1,2,\ldots,n$

$$\dim \Lambda T_m = \binom{n}{m} = \binom{n}{n-m} = \dim \Lambda T_{n-m} \; .$$

In the presence of a <u>volume element</u> on V (that is any

$$e_1 \wedge e_2 \wedge \ldots \wedge e_n \in \Lambda T_n \; ,$$

where the $(e_i)_{i=1,2,\ldots,n}$ constitute an <u>orthonormal</u> basis) there is an isomorphism for each $m = 0,1,2,\ldots,n$

$$* : \Lambda T_m \longrightarrow \Lambda T_{n-m}$$

called the <u>Hodge star operator</u>. In particular

$$* : \Lambda T_0 \longrightarrow \Lambda T_n : a \longmapsto a \, e_1 \wedge e_2 \wedge \ldots \wedge e_n$$

$$* : \Lambda T_n \longrightarrow \Lambda T_0 : a \, e_1 \wedge e_2 \wedge \ldots \wedge e_n \longmapsto a$$

and quite generally $*$ composes with itself as

$$** = (-1)^{m(n-m)} \; I_{\Lambda T_m} \qquad (\forall m = 0,1,2,\ldots,n) \; .$$

Details of the construction can be found in Bishop and Goldberg
[7] p. 107 et seq.   Note that only in the case n = 3 do we have

$$\binom{1}{3} = \binom{2}{3} = 3$$

and consequently the exterior product on $\mathbb{R}^3$ can be interpreted
through the Hodge operator as the vector cross product in $\mathbb{R}^3$,
namely (cf. V §2)

$$v_1 \times v_2 = *(v_1 \wedge v_2) \; .$$

Exterior Analysis

All of the above algebra carries over to the fibres of vector
bundles over a manifold.   Hence we have the space $\Lambda T_r M$ of
antisymmetric $\binom{0}{r}$-tensor fields or (differential) r-forms on M
whose elements are sections of the r-form bundle $\Lambda T_r^0 M$.   Exterior
products for fields are constructed pointwise.

Example 1

Given a chart (U,ϕ) on M then we obtain a frame $(\partial_i)_x$ for
$T_x M$, $x \in U$ and its dual $(dx^i)_x$ for $(T_x M)^*$.   A frame for e.g.
the tensor space $T_2 M_x = (T_x M)^* \otimes (T_x M)^*$ is

$$\{dx^i \otimes dx^j | i,j = 1,2,\ldots,n\} \qquad (cf. \; 1.1)$$

and a frame for the space $\Lambda T_2 M_x$ is

$$\{dx^{i_1} \wedge dx^{i_2} | i_1 < i_2; \; i_1,i_2 = 1,2,\ldots,n\}$$

where by definition

$$dx \wedge dy = \tfrac{1}{2}(dx \otimes dy - dy \otimes dx) \; .$$

The extra ingredient that we have for fields of forms is the
(linear) map exterior differentiation

$$d : \Lambda T_r M \longrightarrow \Lambda T_{r+1} M \; .$$

Amazingly, this very important operator is uniquely characterised
by the following properties (cf. Singer and Thorpe [90] p. 108-111

for proof):

(i)   d is linear.

(ii)   d is the ordinary differential on 0-forms .

(iii) If $\omega \in \Lambda_r TM$, $\sigma \in \Lambda_s TM$ then $d(\omega \wedge \sigma) = d\omega \wedge \sigma + (-1)^r \omega \wedge d\sigma$.

(iv) $d^2 = d \circ d = 0$.

Example 2

Locally $\omega \in \Lambda_r TM$ has an expression

$$\omega = \omega_{(i_1, i_2, \ldots, i_r)} dx^{i_1} \wedge dx^{i_2} \wedge \ldots \wedge dx^{i_r}$$

where the brackets modify the summation convention to act only
over increasing sequences $i_1 < i_2 < \ldots < i_r$.   Then we find
that $d\omega \in \Lambda_{r+1} TM$ has local expression

$$d\omega = d\omega_{(i_1, i_2, \ldots, i_r)} \wedge dx^{i_1} \wedge dx^{i_2} \wedge \ldots \wedge dx^{i_r} .$$

Example 3

In electromagnetic field theory the classical electric and magnetic
vector fields can be represented by one differential 2-form F .
In regions with negligible matter content Maxwell's field equations
are then simply (cf. Clarke [17] and Flanders [33] for more
discussion)

$$dF = 0, \quad *d(*F) = J$$

where J is the electric current 1-form .

Example 4

The composites $*d*$ and $*d$ generalise the classical terms
'divergence' and 'curl' from vector calculus.   There, many familiar
identities are consequences of $d^2 = 0$.

Example 5

If X,Y are vector fields on M and f is a 0-form and ω is
a 2-form then $df(X) = X(f)$

$$d\omega(X,Y) = \tfrac{1}{2}\big(X(\omega(Y))-Y(\omega(X))-\omega[X,Y]\big) \ .$$

Similar formulae obtain for higher forms (cf. Kobayashi and Nomizu [57] p.36).

Grassmann homomorphism

Denote by ΛM the Grassmann algebra of all differential forms on M . Then every smooth map $f : M \longrightarrow N$ between manifolds defines an algebra homomorphism $f^* : \Lambda N \longrightarrow \Lambda M$ by putting, for all $\omega \epsilon \Lambda T_r N$, $f^*\omega \epsilon \Lambda T_r M$ such that

$$f^*\omega(v_1,v_2,\ldots,v_r)_x \ = \ \omega(Dfv_1,Dfv_2,\ldots,Dfv_r)_{f(x)} \ ;$$

and if $r = 0$, $f^*\omega = \omega{\circ}f$.

(What we have here is a contravariant functor at work from Man to the category of Grassmann algebras, cf. II §1.4). It follows that exterior products are preserved and, most significantly, so is exterior differentiation. Precisely we have

(i) $f^*(\omega\wedge\sigma) = f^*\omega\wedge f^*\sigma$

(ii) $f^*d\omega = df^*\omega$.

Proof

Outlined in Singer and Thorpe [90] p. 113 .   ☐

Exactness and closedness

An r-form $\omega\epsilon\Lambda T_r M$ is called:

closed if $d\omega = 0$;

exact if $\exists\sigma\epsilon\Lambda T_{r-1}M : d\sigma = \omega$ (in case $r>0$)

or ω is constant   (in case $r=0$) ;

locally exact if $(\forall x\epsilon M)$ ($\exists$ open neighbourhood U of x) : $\omega|_U$ exact.

We observe the implications (since $d^2 = 0$)

exact $\implies$ locally exact $\implies$ closed .

A deeper consequence is a local converse to

$$\omega = d\sigma \implies d\omega = 0 \qquad (\text{true by } d^2 = 0).$$

Namely: closed differential forms are locally exact.

The proof is outlined in Bishop and Goldberg [7] p. 175 and depends on an intricate homotopy construction.

Vector valued forms

Hitherto our forms have taken values in R , but any vector space V will serve also. A V-valued r-form ω on M is an assignment to each $x \in M$ of a purely antisymmetric r-linear map

$$\omega_x : T_x M \times T_x M \times \ldots \times T_x M \longrightarrow V .$$

Given any frame $(e_1, e_2, \ldots, e_m)$ for V then ω is expressible as a linear combination of (ordinary) r-forms ω^i on M

$$\omega = \omega^1 e_1 + \omega^2 e_2 + \ldots + \omega^m e_m .$$

Then we define the exterior derivative of the V-valued r-form ω to be

$$d\omega = d\omega^1 e_1 + d\omega^2 e_2 + \ldots + d\omega^m e_m .$$

Example 6

The canonical one-form θ of a frame bundle LM is an R^n-valued 1-form (cf. §2.3).

Example 7

The connection form ω of a connection ∇ in a principal fibre bundle (P,G,M) is a ξ-valued 1-form , where ξ is the Lie algebra of G (cf. §2.5,2.6). We know that ξ is essentially the matrix space R^{n^2} if dim M = n and so locally ω is expressible as an n×n array of 1-forms (ω^i_j) with values in R . Suppose that $x \in U$ for some chart (U,ϕ) on M and the mutually dual induced frames for $T_x M$ and $(T_x M)^*$ are $(\partial_i)_x$ and $(dx^i)_x$. Then the connection is locally determined by its components Γ^k_{ij} from

$$\nabla_{\partial_i} \partial_j = \Gamma^\ell_{ij} \partial_\ell \qquad (\text{cf. §2.6}) .$$

Now we find that for each i,j (at $x \in M$)

$$\omega^i_j : T_x M \longrightarrow R : u \longmapsto dx^i(\nabla_u \partial_j) .$$

But ω^i_j is linear so we can characterise it completely by its action on the frame $(\partial_i)_x$:

$$\omega^i_j(\partial_k) = dx^i(\nabla_{\partial_k}\partial_j) = dx^i(\Gamma^\ell_{kj}\partial_\ell) = \Gamma^i_{kj} .$$

Thus, each 1-form ω^i_j admits an expression

$$\omega^i_j = \Gamma^i_{kj}dx^k \epsilon (T_xM)^*$$

and so the connection form ω is expressible as the array

$$\omega = [\omega^i_j] = \begin{pmatrix} \Gamma^1_{k1}dx^k & \Gamma^1_{k2}dx^k & \cdots & \Gamma^1_{kn}dx^k \\ \Gamma^2_{k1}dx^k & \Gamma^2_{k2}dx^k & \cdots & \Gamma^2_{kn}dx^k \\ \vdots & & & \vdots \\ \Gamma^n_{k1}dx^k & \cdots & & \Gamma^n_{kn}dx^k \end{pmatrix}$$

Curvature, torsion and the structure equations

L·t ∇ be a connection in a principal fibre bundle (P,G,M) with connection form ω . We denote the horizontal projection maps induced by ∇ by

$$\hbar : T_uP = H_u \oplus G_u \longrightarrow H_u : X_H \oplus X_G \longrightarrow X_H$$

for all $u \epsilon P$.

The <u>curvature form</u> of ∇ (or of ω) is the 2-form $\Omega = d\omega \circ \hbar$; like ω it takes values in ξ .

The <u>torsion form</u> of ∇ (defined for the case $P = LM$) is the 2-form $\Theta = d\theta \circ \hbar$, where θ is the canonical 1-form, like θ the torsion form takes values in $\mathbb{R}^n$ (cf. §2.3).

The operation on forms σ

$$\sigma \longmapsto d\sigma \circ \hbar$$

is called by Kobayashi and Nomizu [57] p. 77 <u>exterior covariant</u> <u>differentiation</u> and they give details of its properties. The important consequences are as follows, with notation as above:

(i) Ω : $(X,Y) \longmapsto d\omega(X,Y) + \frac{1}{2}[\omega(X),\omega(Y)]$;

(ii) $d\Omega \circ \bar{h} = 0$;

(iii) Θ : $(X,Y) \longmapsto d\theta(X,Y) + \frac{1}{2}\big(\omega(X).\theta(Y) - \omega(Y).\theta(X)\big)$;

(iv) $d\Theta \circ \bar{h} = \Omega \wedge \theta$. $\square$

The results (i) and (iii) are called the structure equations (of
E. Cartan) and (ii) and (iv) are called the Bianchi identities.

Example 8

We have seen that ω can be viewed as an $(n \times n)$-matrix valued form
$[\omega^i_j]$ and similarly Ω is an $(n \times n)$-matrix valued form, $[\Omega^i_j]$ say.
Because θ appears as an $(n \times 1)$-matrix valued form, $[\theta^i]$ say,
Θ is also an $(n \times 1)$-matrix valued form, $[\theta^i]$ say .

Then the structure equations are represented by the equations for
their typical matrix entries:

(i)' $d\omega^i_j = -\omega^i_k \wedge \omega^k_j + \Omega^i_j$;

(iii)' $d\theta^i = -\omega^i_j \wedge \theta^j + \theta^i$.

In this notation one readily obtains an identity sometimes called
the third structural equation,

$$d\Omega^i_j = d\omega^i_k \wedge \omega^k_j - \omega^i_k \wedge d\omega^k_j ,$$

by applying exterior differentiation to (i)' .

Traditionally in spacetime geometry the notion of curvature
was developed via a tensor field, after Riemann. We define this,
and another field, and show how they relate to the 2-forms Ω and
Θ .

The torsion of a connection ∇ in a frame bundle LM is
that $T \in T^1_2 M$ defined by

$$T(X,Y) = \nabla_X Y - \nabla_Y X - [X,Y] , \quad (\forall\ X,Y \in T^1 M)$$

and the curvature of this ∇ in LM is that $R \in T^1_3 M$ defined by

$$R(X,Y)Z = \nabla_X \nabla_Y Z - \nabla_Y \nabla_X Z - \nabla_{[X,Y]} Z ,$$

$$(\forall\ X,Y,Z \in T^1 M) .$$

It follows that any any $x \epsilon M$ with $X, Y, Z \epsilon T_x M$:

$$T(X,Y) \;=\; u(2\Theta(\tilde{X}, \tilde{Y}))$$

$$R(X,Y)Z \;=\; u(2\Omega(\tilde{X}, \tilde{Y}))(u^{-1}Z)$$

where $u \epsilon \overleftarrow{\Pi}_L(x)$ is arbitrary and $\tilde{X}$ is any vector in $T_u LM$ which projects by $D\Pi_L$ onto X etc., and we view $u = (u_i) \epsilon \overleftarrow{\Pi}_L(x)$ as the isomorphism

$$u \;:\; R^n \longrightarrow T_x M \;:\; (a^i) \longmapsto a^i u_i \;.$$

Proofs are given in Kobayashi and Nomizu [57] p. 132 et seq., together with further results.

A connection in LM is called symmetric if its torsion is identically zero.

Example 9

Since $T \epsilon T_2^1 M$ and $R \epsilon T_3^1 M$, they have local expression in terms of the mutually dual frames (∂_i), (dx^i) of the form

$$T \;=\; T_{jk}^i \partial_i \otimes dx^j \otimes dx^k$$

$$R \;=\; R_{jk\ell}^i \partial_i \otimes dx^j \otimes dx^k \otimes dx^\ell$$

where

$$T(\partial_j, \partial_k) \;=\; T_{jk}^i \partial_i$$

and

$$R(\partial_k, \partial_\ell) \partial_j \;=\; R_{jk\ell}^i \partial_i \;.$$

Most books covering elementary differential geometry (e.g. [28]) will show how to deduce the following local expressions for their components in terms of the components of the connection:

$$T_{jk}^i \;=\; \Gamma_{jk}^i - \Gamma_{kj}^i$$

$$R_{jk\ell}^i \;=\; \partial_k \Gamma_{\ell j}^i - \partial_\ell \Gamma_{kj}^i + \Gamma_{kj}^h \Gamma_{\ell h}^i - \Gamma_{\ell j}^h \Gamma_{kh}^i \;.$$

We know already that locally

$$\omega_j^i \;=\; \Gamma_{jk}^i dx^k$$

and applying d , (cf. Example 2 above)

$$d\omega^i_j = d\Gamma^i_{jk} \wedge dx^k$$

$$= \partial_r \Gamma^i_{jk} dx^r \wedge dx^k .$$

Similarly, we have also

$$\omega^i_h \wedge \omega^h_j = \Gamma^i_{hr} dx^r \wedge \Gamma^h_{jk} dx^k$$

$$= \Gamma^i_{hr} \Gamma^h_{jk} dx^r \wedge dx^k$$

and substitution in (i)' yields

$$\Omega^i_j = \left(\partial_r \Gamma^i_{jk} - \Gamma^i_{hr} \Gamma^h_{jk} \right) dx^r \wedge dx^k ,$$

$$= \tfrac{1}{2} R^i_{jrk} \, dx^r \wedge dx^k$$

by the definition of R^i_{jrk} and the exterior product.

Equivalently, using our modified summation convention for increasing sequences of indices :

$$\Omega^i_j = R^i_{j\,(rk)} \, dx^r \wedge dx^k .$$

It follows that we can now write the following formulae (cf. e.g. Dodson [24] for details)

$$\omega^i_j \wedge dx^j = T^i_{(kj)} dx^k \wedge dx^j$$

$$d\omega^i_j = -\omega^i_k \wedge dx^k_j + R^i_{j\,(km)} dx^k \wedge dx^m .$$

It is painfully clear that the actual details of the behaviour of a given connection are rather tedious to calculate. For a manifold of dimension n and a symmetric connection ∇ there are $\tfrac{1}{2} n^2 (n+1)$ independent components Γ^k_{ij} and $n^2(n^2-1)/12$ independent components $R^i_{jk\ell}$ of curvature. On the other hand there are only $\binom{n}{2}$ connection forms ω^i_j and the same number of curvature forms Ω^i_j. In a situation where ∇ has a reasonable amount of symmetry it is possible to solve the first Cartan equation for the ω^i_j by inspection. Then the Ω^i_j are readily calculated and yield the components $R^i_{jk\ell}$ by inspection, thus avoiding their direct calculation via the Γ^k_{ij} and their derivatives from the formula given above. The method is due to Cartan and was highlighted by Misner [71]. We shall employ it at the end of the next section.

Remark

The theory of forms proceeds with their integration, leading to
the (generalized) theorem of Stokes and to the theorems of De Rham
(cf. Bishop and Goldberg [7], Flanders [33] Singer and Thorpe [90],
Spivak [93,94] and Hodge [51].)

The theory of connections is really the main theme in Kobayashi
and Nomizu [57], which is everyone's main reference.   We need one
more notion here, that of a flat connection.

A connection ∇ in a principal fibre bundle (P,G,M) is
called flat if its curvature form vanishes identically.   The
canonical flat connection $\underline{\nabla}$ in $P = M \times G$ is defined by the
decomposition $\forall u = (x,h) \in M \times G = P$

$$T_u (M \times G) = H_u \oplus G_u$$

with $H_u = T_u (M \times \{h\})$.

So ∇ in $M \times G$ is this $\underline{\nabla}$ if and only if it is reducible to the
unique connection in $M \times \{e\}$.

Then it follows that we have

2.7.1   Theorem

(i)   A connection ∇ in a principal fibre bundle (P,G,M) is
flat if and only if every $x \in M$ has an open neighbourhood U such
that there is a bundle isomorphism

$$\psi : \overset{\leftarrow}{\Pi}_p U \longrightarrow U \times G$$

mapping the horizontal subspace at each $u \in \overset{\leftarrow}{\Pi}_p U$ onto the horizontal
subspace at $\psi(u)$ of the canonical flat connection in $U \times G$.

(ii)   If M is paracompact and simply connected and ∇ is a flat
connection in a principal fibre bundle (P,G,M) then P is
isomorphic to the trivial bundle $M \times G$, and the horizontal
subspaces of ∇ are mapped onto those of $\underline{\nabla}$ by this isomorphism.
(Cf. Theorem 3 in §2.3)

Proof

Kobayashi and Nomizu [57] p. 92-3.   □

Corollary 1

A connection ∇ is flat if and only if around every point $x \in M$ there is a neighbourhood N_x such that for any $y \in N_x$ parallel transport gives the same result along any curve in N_x from x to y.

Proof

By (i), if ∇ is flat then each $\overset{\leftarrow}{\Pi}_p U$ admits a section p_U that is parallel with respect to ∇. Locally, the construction resembles that in Theorem 4; namely we choose p_U to satisfy

$$\nabla|_U = \nabla^{p_U} .$$

Then parallel transport is independent of choice of curve in U.

 Conversely, if we have some ∇ for which each point x has a neighbourhood N_x in which parallel transport is independent of curve in N_x, then ∇ determines a section p of $\overset{\leftarrow}{\Pi}_p N_x$. For, choose any $p(x) \in \overset{\leftarrow}{\Pi}_p(x)$ and parallel transport it throughout N_x. Since the map is bijective we obtain a unique $p(y) \in \overset{\leftarrow}{\Pi}_p(y)$ for each $y \in N_x$. This allows us to construct the local bundle isomorphisms, required by the theorem to ensure that ∇ is flat, from

$$\psi : \overset{\leftarrow}{\Pi}_p U \longrightarrow U \times G : R_h p(y) \longmapsto (y,h) . \qquad \square$$

 A direct proof of this corollary for a connection in a frame bundle is given in [28] p. 435 with ample discussion of the geometry involved. Observe that flatness is a local property from either viewpoint: vanishing of curvature differential form or local freedom of parallel transport.

Corollary 2

If ∇ is a flat connection in (P,G,M) over a simply connected manifold M, then parallel transport gives the same result along any curve between each pair of points in M, and P admits a global section.

Proof

If M is simply connected then any two curves from x to y can be continuously deformed one into the other through a family of curves from x to y , i.e. they are homotopic. So we can bridge two such curves by neighbourhoods like N_x , available through Corollary 1 because ∇ is flat.

Parallel transport throughout M yields global sections of P.   □

2.8   Riemannian and pseudo-Riemannian structures

A metric tensor on a vector space V is a symmetric, non-degenerate bilinear map

$$g : V \times V \longrightarrow R : (u,v) \longmapsto u \cdot v .$$

(Recall that g is non-degenerate if

$$\bigl(g(u,v) = 0 \quad (\forall v \in V)\bigr) \Longrightarrow u = \underline{0} .)$$

It matters particularly whether or not a metric tensor is positive definite i.e. $g(u,u) > 0 \quad \forall u \neq 0$.

An inner product is a positive definite metric tensor.
(Of course a negative definite map can always be converted into a positive definite map by a trivial change of sign, without altering the geometry of the space.)

Example 1

The standard inner product on R^n is

$$\cdot : \bigl((x^i),(y^i)\bigr) \longmapsto x^1 y^1 + x^2 y^2 + \ldots + x^n y^n .$$

The Lorentz metric tensor on R^4 is

$$\overset{v}{\cdot} : \bigl((x^i),(y^i)\bigr) \longmapsto -x^1 y^1 + x^2 y^2 + x^3 y^3 + x^4 y^4 ,$$

which is not positive definite. (Sometimes the negative of the given map is called the Lorentz metric tensor.)

A metric tensor field on a manifold M is a section of $T_2 M$, i.e. a $\binom{0}{2}$-tensor field $g \in T_2 M$, such that $(\forall x \in M) \ g_x$ is a metric tensor on $T_x M$.
Such a metric tensor field g is called

(i)   A Riemannian structure on M if $(\forall x \in M) \ g_x$ is an inner product, then (M,g) is a Riemannian manifold ;

(ii) a pseudo-Riemannian structure on M if it is not a
Riemannian structure on M , then (M,g) is a pseudo-Riemannian
manifold.

Manifolds of type (ii) are particularly relevant to spacetime
geometry (cf. V §1, below) but they were sadly neglected by the
earlier pure mathematical texts and indeed still often do not have
due prominence.   The geometry involved in the pseudo-Riemannian
case is often much more intricate; for an introductory treatment
with examples see Dodson and Poston [28] and Bishop and Goldberg
[7] .

Example 2

Euclidean n-space is the Riemannian n-manifold (R^n,g) where, as
always using the identity chart for R^n , $(\forall x \in R^n)$:

$$g_x : T_x R^n \times T_x R^n \longrightarrow R : \big((x,u),(x,v)\big) \longmapsto u \cdot v$$

and $u \cdot v$ denotes the standard inner product on R^n .

Minkowski space is the pseudo-Riemannian 4-manifold (R^4,g)
where $(\forall x \in R^4)$

$$g_x : T_x R^4 \times T_x R^4 \longrightarrow R : \big((x,u),(x,v)\big) \longmapsto u \overset{v}{\cdot} v$$

and $\overset{v}{\cdot}$ is the Lorentz metric tensor on R^4 .

Example 3

The set of unimodular linear operators on R^2 consists of (2×2)-
matrices with determinant 1.   In fact it is also a 3-manifold and
a Lie group (cf. §1.5) , under matrix multiplication; denoted by
$Sl(2;R)$ because it is a subgroup of $Gl(2;R)$ it has a natural
pseudo-Riemannian structure given by

$$g_{\underset{\sim}{x}} : \left(\big(\underset{\sim}{x} , \begin{bmatrix} a & b \\ c & d \end{bmatrix}\big) , \big(\underset{\sim}{x} , \begin{bmatrix} p & q \\ r & s \end{bmatrix}\big)\right) \longmapsto \tfrac{1}{2}(as{+}dp) - \tfrac{1}{2}(br{+}cq)$$

where we use the identity chart for $Sl(2;R)$ as a subset of R^4 .
The geometry of this example is discussed at length in [28]
p. 394 et seq.

Example 4

The lecture notes of MacLane [66] on geometrical mechanics develop

the theme that, for a physical system, **kinetic energy** is a
Riemannian structure on configuration space.

Example 5

If M is parallelizable by

$$p : M \longrightarrow LM : x \longmapsto (p_i)_x$$

then M admits a Riemannian structure g_p by putting

$$g_p : T_xM \times T_xM \longrightarrow \mathbb{R} : (a^i p_i, b^j p_j) \longmapsto a^i b^j \delta_{ij}$$

where $\delta_{ij} = 1$ if $i = j$ else $\delta_{ij} = 0$.   Further consequences
are given below in Theorem 6 and ⁺s corollaries.

   Next we collect some theorem on metric structures.

2.8.1   Theorem

A paracompact manifold M always admits a Riemannian structure.

Proof

A partition of unity allows a smoothing out of a local structure -
Brickell and Clark [9] p. 163.   The construction fails to adapt
to produce indefinite metric tensor fields.   In fact the torus
and the Klein bottle are the **only** compact (Hausdorff) 2-manifolds
to admit a pseudo-Riemannian structure (cf. Steenrod [96]
p. 207).   □

Corollary

In particular the result holds for connected M with a countable
basis for its topology, since such M have partitions of unity.   □

2.8.2   Theorem

A metric tensor field g on a manifold M determines a unique
connection ∇^g in LM such that

(i) ∇^g is symmetric (i.e. has torsion zero)

(ii) ∇^g is **compatible with** g (i.e. parallel transport of
tangent vectors is always an isometry;   this is equivalent to
requiring :

$$u(g(v,w)) = g(\nabla^g_u v, w) + g(v, \nabla^g_u w), \quad \forall u,v,w \in T^1M).$$

Proof

Details are given in e.g. [28] p. 328.   Locally, the components
Γ^k_{ij} of ∇^g are determined by the components $g(\partial_i, \partial_j) = g_{ij}$ of
g by :

$$\Gamma^k_{ij} = \tfrac{1}{2} g^{km} (\partial_i g_{jm} + \partial_j g_{im} - \partial_m g_{ij})$$

where g^{km} denotes the km-th component of the inverse of $[g_{ij}]$.
The connection so determined is called the **Levi-Civita** or **metric**
connection of g .   □

2.8.3   Theorem

Let (M,g) be a connected Riemannian manifold then M is
metrizable because we have:

(i)   a topological **metric** or **distance function**

$$d_g : M \times M \longrightarrow R$$

$$: (x,y) \longmapsto \inf \left\{ \int \| \dot{c} \| \; \Big| \; c \text{ is a curve from } x \text{ to } y \right\}.$$

(ii) The topology determined by d coincides with the manifold
topology on M .

Proof

Details are provided e.g. in Kobayashi and Nomizu [57] p. 157-166.
Note that here 'curve' means piecewise C^1-curve and that the
integral is the **length** of the curve i.e.

$$\int \| \dot{c} \| = \int_0^1 \Big(g_{c(t)} \big(\dot{c}(t), \dot{c}(t) \big) \Big)^{\tfrac{1}{2}} dt ,$$

where $c : [0,1] \longrightarrow M$ with $c(0) = x$, $c(1) = y$.   □

Corollary 1

If (M, d_g) is not **complete** as a metric space then it can be
completed by the standard procedure (cf. e.g., Maddox [68]) of
supplying limits for its Cauchy sequences.   □

Corollary 2

Like all spaces with a metric topology, M is paracompact.   □

We say that a Riemannian manifold (M,g) is <u>complete</u> if the associated metric space (M,d_g) is complete in the sense that every Cauchy sequence is convergent.

2.8.4   <u>Theorem</u>

If (M,g) is a connected Riemannian manifold then

(i) (M,g) is complete if and only if ∇^g is <u>geodesically complete</u> (i.e. every geodesic can be extended to domain $\mathbb{R}$ for an affine parameter, such as arc length).   This fails for pseudo-Riemannian g (cf. V §6 below).

(ii) (M,g) is complete if and only if every d_g-bounded subset of M is relatively compact.

(iii) If ∇^g is geodesically complete then any two points x,y in M can be joined by a minimizing geodesic; for x,y close enough this geodesic is unique.

(iv)   If all geodesics starting from any particular point of M are infinitely extensible then ∇^g is geodesically complete.

(v) (M,g) is complete if M is compact.

(vi) (M,g) is complete if its group of <u>isometries</u> (g-preserving diffeomorphisms) is transitive on M .

<u>Proof</u>

Kobayashi and Nomizu [57] p. 172 et seq.   □

The next result is much newer, apparently discovered independently by Schmidt [84] and Marathe [69], though possibly known to Ehresmann.

2.8.5   <u>Theorem</u>

Let H be the closed subgroup of $G\ell(n;\mathbb{R})$ which leaves invariant a given metric tensor on $\mathbb{R}^n$.   Given an n-manifold M denote by E the associated bundle of LM with fibre G/H .   Then M is paracompact if E admits a section.

<u>Proof</u>

This is due to Marathe [69] using the following constructions.

(i)   Each section $\sigma : M \longrightarrow E$ determines (cf. §2.3 Theorem 2)
a unique reduced subbundle P_σ of LM with structure group H .

(ii) A metric tensor on $\mathbb{R}^n$ induces a unique symmetric connection
in P_σ and provides M with a pseudo-Riemannian structure g .

(iii) (M,g) determines the Levi-Civita connection ∇^g in LM and
hence a Riemannian structure

$$\hat{g} : TLM \times TLM \longrightarrow \mathbb{R}$$

$$: (X,Y) \longmapsto \theta(X) \cdot \theta(Y) + \omega(X) \cdot \omega(Y) ,$$

where θ is the canonical 1-form , ω is the connection form of
∇^g and $\cdot$ is the standard inner product on $\mathbb{R}^n$ and $\mathbb{R}^{n^2}$.   (We
view the Lie algebra ξ as the matrix space $\mathbb{R}^{n^2}$.)

(iv) (LM,$\hat{g}$) is metrizable and hence paracompact.   Each component
of LM is locally compact and paracompact, so can be written as a
countable union of nested compact sets K_n with each K_n contained
in the interior of K_{n+1} .   Each K_n is metrizable and therefore
so is each $\Pi_L(K_n) \subset M$ metrizable.   Consequently, M is metrizable
and thus paracompact.   □

Corollary

M is paracompact if and only if LM admits a connection; in
particular, M is paracompact if and only if it admits a
Riemannian structure.

Proof

This result was also known to Geroch [37] who proved that if M
admits a connection (class C^1 will do, since it is existence and
openness of the exponential map that is used in his construction)
then M has a countable basis for its topology and is therefore
paracompact.   The converse is obtainable from our Theorems 1
and 2 .   □

The metric $\hat{g}$ induced in the frame bundle LM by a connection
∇ is useful in the definition of spacetime singularities, as we
shall see in Ch. V §5.   In the presence of more structure (a
parallelization p) it is possible to achieve an improved
resolution of these singularities.   The technique is to embed a

manifold M in $(LM, \tilde{g}_p)$ where $\tilde{g}_p = \hat{g} + \tilde{p} \cdot \tilde{p}$ i.e.

$$\tilde{g}_p : TLM \times TLM \longrightarrow R$$

$$: (X,Y) \longmapsto \theta(X) \cdot \theta(Y) + \omega(X) \cdot \omega(Y) + \tilde{p}(X) \cdot \tilde{p}(Y)$$

and $\tilde{p}$ is the connection form induced by the connection ∇^p
associated with the parallelization p (cf. Theorem 4 §2.6).

2.8.6   Theorem

Every parallelization p on a connected manifold M determines a
unique Riemannian structure g_p on M and a unique Lorentz
structure on M .

Proof

The parallelization

$$p : M \longrightarrow LM : x \longmapsto (p_i)_x$$

determines at each $x \in M$ an isomorphism

$$T_x M \equiv R^n : a^i p_i \longmapsto (a^i) ,$$

which we can make into an isometry by putting

$$g_p(x) : T_x M \times T_x M \longrightarrow R$$

$$: (a^i p_i, b^j p_j) \longmapsto (a^i) \cdot (b^j) .$$

So we are using the standard inner product $\cdot$ on R^n to generate
via p the Riemannian structure on M :

$$g_p : M \longrightarrow T_2 M : x \longmapsto g_p(x) .$$

Since LM consists of <u>ordered</u> bases, we can replace $\cdot$ by the
Lorentz structure on R^4 . $\Box$

Corollary 1

By Theorem 3, g_p determines a unique topological metric d_p on
M and this metric topology coincides with the manifold topology,
so M is paracompact. $\Box$

Corollary 2

By Theorem 2, g_p uniquely determines the Levi-Civita connection
∇^{g_p} and its connection form ω_p .   Hence, by Theorem 5, LM has a
Riemannian structure

$$\hat{g}_p = \theta \cdot \theta + \omega_p \cdot \omega_p$$

and is therefore paracompact.   □

Corollary 3

By construction : for all $x \epsilon M$, if $v \epsilon T_x M$ then $D_x p(v)$ is horizontal with respect to ∇^p and so

$$\tilde{p}(D_x p(v)) = O, \quad (\forall v \epsilon T_x M) .$$

Hence, $(\forall v, w \epsilon T_x M)$, with $\bar{g}_p = \theta \cdot \theta + \tilde{p} \cdot \tilde{p}$ we have

$$\bar{g}_p (D_x p(v), D_x p(w)) = \theta D_x p(v) \cdot \theta D_x p(w) ,$$

which, in general, differs from $g_p(v,w)$ so the parallelization is not necessarily an isometry.   □

Corollary 4

Any parallelization q that is a section of the orthonormal bundle, determined by the metric g_p of a parallelization p , gives a connection ∇^q that (like ∇^p) is compatible with g_p .

Proof

By hypothesis, q gives at each $x \epsilon M$ an orthonormal frame $(q_i)_x$ and so

$$q_k g_p (q_i, q_j) = O \quad (\forall i, j, k) .$$

But also ∇^q satisfies

$$\nabla^q_{q_i} q_j = \underline{O} \quad (\forall i, j) .$$

So the compatibility condition is satisfied identically.   □

We have seen how a parallelization determines a connection, the reverse process also works for simple connections.   The idea is to choose a frame at some point and then to determine a frame at other points by parallel transport of the starting frame. Plainly, this will only determine a parallelization if, at least locally, the result of parallel transport does not depend on the choice of curve.   So what we need is a flat connection, that is one with zero curvature (discussed at the end of §2.8).

2.8.7   Theorem

(i)   Any flat connection ∇ in a frame bundle LM over simply
connected M determines a parallelization p that satisfies
$\nabla^p = \nabla$; moreover, every parallelization connection is flat.

(ii)   If a metric tensor field g on a manifold M determines a
non-flat Levi-Civita connection ∇^g and M has a parallelization
p that is orthonormal in g , then the connection ∇^p is not
symmetric.

Proof

(i)   Use the Theorem at the end of §2.8.   The flatness of ∇
allows us to bring back to LM the parallelization M×{e} of
M×Gℓ(n;ℝ) corresponding to the canonical flat connection.   We
exploit simple connectedness and use the local bundle isomorphisms.
Hence we obtain a parallelization of LM which therefore determines
a connection ∇^p .   But by construction, the horizontal subspaces
defined by ∇^p coincide with those of ∇ .   So $\nabla^p = \nabla$.

Any parallelization connection ∇^p is flat by virtue of the
same Theorem as in (i).   For p stratifies LM globally and
hence makes it isomorphic to M×Gℓ(n;ℝ) , mapping horizontal
subspaces of ∇^p onto those of the canonical flat connection.

(ii)   If M has a parallelization p then we know by (i) that
∇^p is flat.   Also, LM is trivial and so in the presence of a
metric tensor field g we can choose p to lie in a connected
component of the orthonormal frame bundle.   Hence the compatibility
equation is again satisfied by both sides being zero, so ∇^p is
compatible with g .

Now, ∇^g is uniquely determined by symmetry and compatibility
with g .   Hence, if ∇^g is not flat then it must differ from
∇^p which is flat.   Therefore ∇^p is not symmetric.   □

In Theorem 9 below we shall see that ∇^p is always
compatible with g_p , precisely because p is orthonormal in g_p
and parallel with respect to ∇^p .   We also work out there, in an
example, the components of ∇^p and ∇^{g_p} , displaying their
compatibilities with g_p .

We saw that a parallelization p uniquely determines a

Riemannian structure g_p on M . This can be useful when (M,g)
is a pseudo-Riemannian manifold, for then we have many choices for
a Riemannian structure via a partition of unity and a parallelization
can help to choose among them. However, if we still wish to make
direct use of a given pseudo-Riemannian structure in constructing
a Riemannian structure we can do so via a parallelization, as we
see next.

2.8.8   Theorem

Let (M,g) be a connected pseudo-Riemannian manifold with a
parallelization p . Then there is induced on M a Riemannian
structure $\hat{p}$, through the inclusion of pM in the Riemannian
manifold $(LM,\hat{g})$.

Proof

Evidently p is a diffeomorphism of M onto pM and so pM is
actually a submanifold of LM . Hence $(pM, \hat{g}|_{pM})$ is a Riemannian
manifold with $pM \equiv M$ and so a Riemannian structure on M is
given by putting

$$\hat{p}(v,w) = \hat{g}(D_x p(v), D_x p(w)) \qquad (\forall v, w \in T_x M) . \qquad \square$$

Corollary

Riemannian structures on M are apparently given by isometrically
embedding $M \equiv pM$ in :

(i) $(LM,\tilde{g}_p)$ where $\tilde{g}_p = \hat{g} + \tilde{p} \cdot \tilde{p} = \theta \cdot \theta + \omega \cdot \omega + \tilde{p} \cdot \tilde{p}$,

(ii) $(LM,\hat{g}_p)$ where $\hat{g}_p = \theta \cdot \theta + \omega_p \cdot \omega_p$,

(iii) $(LM,\bar{g}_p)$ where $\bar{g}_p = \theta \cdot \theta + \tilde{p} \cdot \tilde{p}$.

However, on pM we have

$$\tilde{g}_{p|pM} = \hat{g}|_{pM}$$

so the same metric structure is induced on M . Also, $\hat{g}_p$ is a
special case of $\hat{g}$, and the parallelization metric g_p on M
coincides with that induced by the restriction of $\bar{g}_p$ to pM . $\square$

Of these, only case (i) has dependence on the given pseudo-
Riemannian structure, via its Levi-Civita connection form ω .

Also, cases (ii) and (iii) are qualitatively different because ω_p is the Levi-Civita connection form of the parallelization metric structure on M but $\tilde{p}$ is the connection form of the parallelization connection. Our next theorem compares the two connections ∇^p and ∇^{g_p} .

2.8.9   Theorem

A parallelization connection ∇^p is necessarily compatible with the parallelization metric g_p , however it need not be symmetric and therefore need not be the Levi-Civita connection of any Riemannian or pseudo-Riemannian structure.   In particular, ∇^p may differ from the Levi-Civita connection ∇^{g_p} .

Proof

We show compatibility directly then the failure of symmetry in a simple example.

(i)   The definition of ∇^p for a parallelization

$$p : M \longrightarrow LM : x \longmapsto (p_i)_x$$

is :

$$\nabla^p_{p_i} p_j = \underline{0} \qquad \forall i,j .$$

The metric tensor field g_p satisfies (cf. Theorem 7)

$$g_p(p_i,p_j) = 1 \text{ if } i=j, \quad 0 \text{ if } i \neq j .$$

Hence, each p_j is a parallel unit vector field and the p_i are mutually orthogonal in g_p .   In particular, the requirement for compatibility is

$$p_k(g_p(p_i,p_j)) = g_p(\nabla^p_{p_k} p_i,p_j) + g_p(p_i,\nabla^p_{p_k} p_j) .$$

This is its formulation via the Ricci Lemma : that g_p have vanishing covariant derivative (cf. [28] p. 340).   It is met identically by both sides vanishing in our case.

(ii) To show that ∇^p need not be symmetric we shall construct a simple counterexample using :

$$p : \mathbb{R}^2 \longrightarrow L\mathbb{R}^2 : (x,y) \longmapsto (e^x \partial_1, e^x \partial_2)$$

where (∂_1, ∂_2) is the standard frame for $T_{(x,y)}R^2$, via the identity chart on R^2.

This p is a parallelization and it determines a connection ∇^p by the conditions (cf. Theorem 4 §2.6):

$$\nabla^p_{\partial_i} e^x \partial_j = \underline{0} \qquad \text{(for } i,j = 1,2\text{)}$$

where we have used the fact that e^x is never zero, and for any connection ∇, $\nabla_v w$ is linear in v.

Expanding this we find

$$e^x \partial_j + e^x \Gamma^k_{1j} \partial_k = \underline{0} \qquad \text{(for } i = 1\text{)}$$

$$e^x \Gamma^k_{2j} \partial_k = \underline{0} \qquad \text{(for } i = 2\text{)}.$$

Then we can deduce the two matrices of components :

$$[\Gamma^1_{ij}] = \begin{bmatrix} -1 & 0 \\ 0 & 0 \end{bmatrix} \quad \text{and} \quad [\Gamma^2_{ij}] = \begin{bmatrix} 0 & -1 \\ 0 & 0 \end{bmatrix}.$$

So ∇^p fails to be symmetric because

$$\Gamma^2_{12} \neq \Gamma^2_{21}.$$

Next, p determines a Riemannian structure g_p (via Theorem 6 above) which in standard coordinates has component matrix at $(x,y) \in R^2$ given by

$$[g_{ij}] = \begin{bmatrix} e^{-2x} & 0 \\ 0 & e^{-2x} \end{bmatrix}.$$

In order to check whether a given connection ∇ is compatible with a given metric tensor field g we can employ Ricci's Lemma (cf. [28] p. 340). It is equivalent to satisfaction of the equation

$$u(g(v,w)) = g(\nabla_u v, w) + g(v, \nabla_u w)$$

for all tangent vector fields u,v,w. We shall use the component form with $u = \partial_k$, $v = \partial_i$, $w = \partial_j$. Taking $g = g_p$, the left hand side is given by the pair of matrices

$$[\partial_1 g_{ij}] = \begin{bmatrix} -2e^{-2x} & 0 \\ 0 & -2e^{-2x} \end{bmatrix}, \quad \partial_2 g_{ij} = \begin{bmatrix} 0 & 0 \\ 0 & 0 \end{bmatrix}.$$

Taking $\nabla = \nabla^p$, the right hand side of the equation is

$$g_p(\Gamma^m_{ki}\partial_m, \partial_j) + g_p(\partial_i, \Gamma^m_{kj}\partial_m)$$

$$= \quad g_{mj}\,\Gamma^m_{ki} + g_{mi}\Gamma^m_{kj}$$

$$= \quad e^{-2x}\Gamma^j_{ki} + e^{-2x}\Gamma^i_{kj}\,.$$

From above :

$$[\Gamma^j_{1i}] = \begin{bmatrix} -1 & 0 \\ 0 & -1 \end{bmatrix}, \quad [\Gamma^j_{2i}] = \begin{bmatrix} 0 & 0 \\ 0 & 0 \end{bmatrix}$$

$$[\Gamma^i_{1j}] = \begin{bmatrix} -1 & 0 \\ 0 & -1 \end{bmatrix}, \quad [\Gamma^i_{2j}] = \begin{bmatrix} 0 & 0 \\ 0 & 0 \end{bmatrix}.$$

By adding these matrices we see that ∇^p is indeed compatible with g_p but since it is not symmetric it is not the Levi-Civita connection of <u>any</u> metric tensor field.

In fact, the Levi-Civita connection ∇^{g_p} determined by the parallelization metric g_p can be found by solving the equation of Ricci's Lemma. In coo:dinates then, we seek those components $\tilde{\Gamma}^k_{ij}$ satisfying

$$\nabla^{g_p}_{\partial_i}\partial_j = \tilde{\Gamma}^k_{ij}\partial_k \qquad \text{(definition of } \nabla^{g_p}\text{)}$$

$$g_{km}\tilde{\Gamma}^k_{ij} = \tfrac{1}{2}(\partial_i g_{jm} + \partial_j g_{im} - \partial_m g_{ij}) \qquad \text{(Ricci)}\,.$$

It turns out that they are given by

$$[\tilde{\Gamma}^1_{ij}] = \begin{bmatrix} -1 & 0 \\ 0 & 1 \end{bmatrix}, \quad \tilde{\Gamma}^2_{ij} = \begin{bmatrix} 0 & -1 \\ -1 & 0 \end{bmatrix}.$$

Symmetry is evident and compatibility with g_p is displayed by

$$[\tilde{\Gamma}^j_{1i}] = \begin{bmatrix} -1 & 0 \\ 0 & -1 \end{bmatrix}, \quad [\tilde{\Gamma}^j_{2i}] = \begin{bmatrix} 0 & -1 \\ 1 & 0 \end{bmatrix}$$

$$[\tilde{\Gamma}^i_{1j}] = \begin{bmatrix} -1 & 0 \\ 0 & -1 \end{bmatrix}, \quad [\tilde{\Gamma}^i_{2j}] = \begin{bmatrix} 0 & 1 \\ -1 & 0 \end{bmatrix},$$

for by summing we see that for $k = 1,2$

$$[\partial_k g_{ij}] = e^{-2x}[\tilde{\Gamma}^j_{ki} + \tilde{\Gamma}^i_{kj}]\,. \qquad\qquad \square$$

Example

Calculation of Schwarzschild geometry using the Cartan structure equations

This particular example of a pseudo-Riemannian 4-manifold is amply discussed in books on relativity, e.g. Misner, Thorne and Wheeler [72], Clarke [17], Hawking and Ellis [43], Dodson and Poston [28]. Here we set out to find (M,g) where M is <u>topologically</u> R^4 and we suppose it has a global chart yielding coordinates $(x^i) = (r,\theta,\phi,t)$ where $t \in R$ is time and $(r,\theta,\phi) \in R^3$ are (pseudo) spherical polar coordinates, valid for $r > 2m$ where m is a positive number yet to be fixed. In fact, (M,g) can successfully represent the geometry of our universe in a region devoid of matter, external to a 'spherically symmetric' mass m and 'static' in time. This can be achieved by seeking that g for which the Ricci tensor vanishes, among the family of Lorentz type metric tensor fields having components

$$g_{11} = f^{-2}, \; g_{22} = r^2, \; g_{33} = r^2\sin^2\theta, \; g_{44} = -f^2$$

where f is a real positive function depending on coordinate r only. The normal boundary conditions are $f(r) \longrightarrow 1$ as $r \longrightarrow \infty$, to represent asymptotic flatness of the geometry at large distances from the central mass. For, in the above coordinates the standard Lorentz structure $\bar{g}$ on flat spacetime (Minkowski space) is given by

$$\bar{g}_{11} = 1, \; \bar{g}_{22} = r^2, \; \bar{g}_{33} = r^2\sin^2\theta, \; g_{44} = -1 \; .$$

The most efficient use of the Cartan equations is via an orthonormal basis field of 1-forms. Such a basis is, by inspection,

$$(\omega^i) \;\; = \;\; (f^{-1}dr, \; rd\theta, \; r\sin\theta \, d\phi, \; f \, dt) \; .$$

Now we proceed as follows:

(i)   find each $d\omega^i$ and identify each connection form ω^i_j in the first Cartan equation ;

(ii)   find the curvature forms by exterior differentiation ;

(iii) extract the $R^i_{jk\ell}$ by inspection ;

(iv)   find the components $R_{jk} = R^i_{jki}$ of the Ricci tensor.

Step (i)

$d\omega^1 = d(f^{-1}) \wedge dr = 0 \; ; \quad dr = f \, \omega^1$.

$d\omega^2 = dr \wedge d\theta \quad = (f/r) \, \omega^1 \wedge \omega^2$.

$d\omega^3 = \sin\theta \, dr \wedge d\phi + \cos\theta \, d\theta \wedge d\phi$.

$\quad = (f/r) \, \omega^1 \wedge \omega^3 + (\cot\theta)/r \, \omega^2 \wedge \omega^3$.

$d\omega^4 = f'dr \wedge dt = f' \, \omega^1 \wedge \omega^4 \; ; \quad f' = \dfrac{df}{dr}$.

The first Cartan equation, $d\omega^i = -\omega^i_j \wedge \omega^j$, gives:

$d\omega^4 = -\omega^4_1 \wedge \omega^1 - \omega^4_2 \wedge \omega^2 - \omega^4_3 \wedge \omega^3 \implies \omega^4_1 = f' \, \omega^4$.

$d\omega^2 = -\omega^2_1 \wedge \omega^1 - \omega^2_3 \wedge \omega^3 - \omega^2_4 \wedge \omega^4 \implies \omega^2_1 = (f/r)\omega^2$.

$d\omega^3 = -\omega^3_1 \wedge \omega^1 - \omega^3_2 \wedge \omega^2 - \omega^3_4 \wedge \omega^4 \implies \omega^3_1 = (f/r)\omega^3$,

$$\omega^3_2 = (\cot\theta)/r \; \omega^3, \omega^3_4 = 0.$$

Now, since (ω^i) is an orthonormal basis at each point then so is its dual (b_i) , hence $g_{ii} = g(b_i,b_i)$ satisfies:

$$g_{11} = g_{22} = g_{33} = 1 \; , \; g_{44} = -1 \; .$$

It follows that since, $\omega^i_j = \Gamma^i_{kj} \, \omega^k$, then

$$\mathrm{sgn}(g_{ii})\omega^i_j = - \, \mathrm{sgn}(g_{jj})\omega^j_i \; .$$

Hence we deduce that $\omega^3_4 = 0$ implies $\omega^4_2 = 0$ in $d\omega^4$.

The collected results so far are:

$\omega^2_1 = -\omega^1_2 = (f/r)\omega^2 \quad = fd\theta$,

$\omega^3_1 = -\omega^1_3 = (f/r)\,\omega^3 \quad = f\sin\theta \, d\phi$,

$\omega^4_1 = \omega^1_4 = f'\omega^4 \quad\quad = ff' \, dt$,

$\omega^3_2 = -\omega^2_3 = (\cot\theta)/r \; \omega^3 = \cos\theta \, d\phi$,

$\omega^4_2 = \omega^2_4 = \omega^4_3 = \omega^3_4 = 0$.

(Note that we could easily extract the Γ^i_{kj} from these equations.)

Step (ii)

We apply exterior differentiation to the ω^i_j :

$$d\omega^2_1 \;=\; df \wedge d\theta \;\;=\;\; (ff')/r \; \omega^1 \wedge \omega^2 \;,$$

$$d\omega^3_1 \;=\; (ff')/r\omega^1 \wedge \omega^2 + (f/r)\cot\theta \; \omega^2 \wedge \omega^3 \;,$$

$$d\omega^4_1 \;=\; \left((f')^2 + ff''\right) \omega^1 \wedge \omega^4 \;,$$

$$d\omega^3_2 \;=\; (-1/r^2) \; \omega^2 \wedge \omega^3 .$$

The second Cartan equation, $\Omega^i_j \;=\; d\omega^i_j + \omega^i_k \wedge \omega^k_j$, gives us the curvature forms by inspection:

$$\Omega^4_1 \;=\; d\omega^4_1 + \omega^4_2 \wedge \omega^2_1 + \omega^4_3 \wedge \omega^3_1 \;=\; \left(ff' + (f')^2\right) \omega^1 \wedge \omega^4$$

$$\Omega^4_2 \;=\; d\omega^4_2 + \omega^4_1 \wedge \omega^1_2 + \omega^4_3 \wedge \omega^3_2 \;=\; \left(ff'/r\right) \omega^3 \wedge \omega^4$$

$$\Omega^3_1 \;=\; d\omega^3_1 + \omega^3_2 \wedge \omega^2_1 + \omega^3_4 \wedge \omega^4_1$$
$$\;=\; \left(ff'/r\right) \omega^1 \wedge \omega^3 + \left(f/r^2\right) \cot\theta \; \omega^2 \wedge \omega^3$$

$$\Omega^2_1 \;=\; d\omega^2_1 + \omega^2_3 \wedge \omega^3_1 + \omega^2_4 \wedge \omega^4_1 \;=\; \left(ff'/r\right) \omega^1 \wedge \omega^2$$

$$\Omega^3_2 \;=\; d\omega^3_2 + \omega^3_1 \wedge \omega^1_2 + \omega^3_4 \wedge \omega^4_2 \;=\; \left(f^2-1\right)/r^2 \; \omega^2 \wedge \omega^3$$

$$\Omega^4_3 \;=\; d\omega^4_3 + \omega^4_1 \wedge \omega^1_3 + \omega^4_2 \wedge \omega^2_3 \;=\; \left(ff'/r\right) \omega^3 \wedge \omega^4 .$$

Step (iii)

Next we use $\Omega^k_1 = R^k_{1(ij)} \; \omega^i \wedge \omega^j$, with summation indicated over increasing ij .

Hence, by inspection from Ω^4_1 ,

$$R^4_{114} \;=\; ff' + (f')^2$$

$$R^4_{1ij} \;=\; 0 \;\; \text{for other} \; i \leqslant j .$$

Similarly,

$$R^3_{113} \;=\; (ff'/r) \;=\; R^2_{112} \;\; \text{and} \;\; R^k_{1ik} = 0 \;\; \text{for} \; i \neq 1, \; \forall k;$$

$$R^1_{441} \;=\; R^4_{114} \;\;=\; ff'' + (f')^2 \;,$$

$$R^2_{442} \;=\; R^4_{224} \;=\; R^4_{334} \;=\; R^3_{443} \;=\; (ff'/r) \;,$$

$$R^1_{331} \;=\; R^3_{113} \;\;=\; (ff'/r) \;,$$

$$R^2_{332} \;=\; R^3_{223} \;\;=\; (f^2/r^2) - (1/r^2) \;,$$

$$R^1_{221} \ = \ R^2_{112} \ = \ (ff'/r) \ .$$

Step (iv)

From $R_{jk} = R^i_{jki}$ we find that the only surviving terms are the diagonal elements:

$$R_{11} \ = \ ff'' + (f')^2 + 2ff'/r$$

$$R_{22} \ = \ R_{33} \ = \ f^2/r^2 + 2ff'/r - 1/r^2$$

$$R_{44} \ = \ ff' + (f')^2 + 2ff'/r \ .$$

Finally, in our particular situation the object was to find the function f such that $[R_{jk}]$ is identically zero.   Setting each R_{ii} to zero we find two equations

$$(f^2/r^2) + (2ff'/r) - (1/r^2) \ = \ 0 \ ,$$

$$ff' \qquad + (f')^2 \qquad + (2ff'/r) = 0 \ .$$

However, the first of these implies the second by differentiation, and we can write it in the form

$$f^2 + (f^2)'r - 1 \ = \ 0 \qquad \text{since} \quad r > 2m > 0 \ .$$

The particular solution that fits observations around the sun is

$$f(r) \ = \ \left(1 - (2m/r)\right)^{\frac{1}{2}} \ ,$$

and we see that the curvature components are dominated for small r by $1/r^3$.

V Spacetime structure

The most important reference work on the geometrical physics of
spacetime is that of Hawking and Ellis [43]. Our present chapter
is intended to complement that work by gathering together various
topological results and by presenting the current position on
singularities.

By a spacetime we shall mean a connected, non-compact Hausdorff
space M with a C^∞ real 4-manifold structure and a C^∞ metric
tensor field g of Lorentz type - the latter means that at each
$x \in M$ the metric tensor on $T_x M$ is representable as a quadratic
form on $\mathbb{R}^4$ with eigenvalues $(-1,1,1,1)$. (See Clarke [13] for
the important embedding theorem.)

We shall see that the existence of a Lorentz structure
guarantees paracompactness (since M is Hausdorff) and conversely
on any non-compact, paracompact Hausdorff manifold there exists a
Lorentz structure that does not generate closed timelike curves.
It is known that there is a Lorentz type structure on a compact
paracompact Hausdorff manifold if and only if the Euler
characteristic is zero, but compactness is unphysical in the sense
that it leads to closed timelike curves - violating accepted rules
of causality. We may as well assume that M is connected since
disconnected parts of a universe would presumably not interact.
The smoothness of M and g is largely for convenience; indeed,
as we pointed out in [28], the bigger step for manifold structures
is from C^0 to C^1 , rather than from C^1 to C^∞ . However,
we do need g to determine a connection and hence geodesics,
covariant derivatives, exponential maps and parallel transport.
The existence theorem at work in each case is essentially the

classical Cauchy Theorem for ordinary differential equations
(cf. [28] p. 276 and Lang [60]).   This assures us of a unique
(local) solution if the function to be integrated satisfies a
Lipschitz Condition on some open set U ; then, moreover, the
solution depends continuously on the initial conditions.   In our
situation the function to be integrated will depend on the
components Γ^k_{ij} of the Levi-Cività connection induced by g (cf.
§2.6), and U will be a subset of R^4 . A sufficient (but not
necessary) requirement for the Lipschitz Condition is that the
first partial derivatives of the Γ^k_{ij} exist and are bounded on a
convex U .   Now, we have seen that the Γ^k_{ij} are determined by
the first partial derivatives of the components g_{ij} of the metric
tensor field g .   So to apply the existence and uniqueness theorem
(to geodesics, parallel transport etc.) we need existing and
bounded second partial derivatives of the g_{ij} .   That is, g
must be of class C^{2-} .   Now, in order to support a C^{2-} field
the differentiable structure on M must be of class C^3 .   On the
other hand it is known (cf. Munkres [73]) that if $r \geqslant 1$ then any
C^r atlas for M has a C^∞ subatlas, so our resort to the
convenience of a C^∞ structure for M is not a serious restriction
for most purposes.

We shall suppose that (M,g) is <u>inextensible</u> as a C^∞ Lorentz
4-manifold; that is to say it is not an isometric proper
submanifold of any other C^∞ Lorentz 4-manifold (cf. Hawking
and Ellis [43] p. 58).   The motive for this supposition is that
no physically reasonable spacetime should be the result of an
arbitrary removal of certain points from a larger spacetime.

We have not made any preparation to study the algebraic
topology of spacetime in this book but recent studies in this field
have been made, for example by Lee [64] and Whiston [102, 103] .
Also, Clarke [14] and Ihrig [53] have used spacetime holonomy
groups and much general background material will be found in [22,
23] .   We shall not here put jet bundles as such to work, but
these will no doubt gain in importance for physical applications
(cf. Hennig [44] and Hermann [45]).

The most important Lie group that arises in spacetime theory
is the <u>Lorentz group</u> O(1,3) .   A thorough study of its

representations and physical applications can be found in Gel'fand,
Minlos and Shapiro [35], an introduction to its role in relativity
is provided in Clarke [17], and Porteous [78] shows how it fits
into the general pattern of transformation groups.   We collect
here a few properties for easy reference.

(i)   As a matrix group, we have the representation

$$O(1,3) = \{\lambda \in Gl(4;R) \mid \lambda^T g_o \lambda = g_o\}$$

where λ^T denotes the transpose of λ and g_o is the diagonal
matrix with elements $(-1,1,1,1)$; so g_o is actually the
quadratic form on R^4 that represents the Lorentz metric tensor
(cf. IV §2.8).   It follows from the product rule for determinants
that

$$\lambda \in O(1,3) \implies \det \lambda = \pm 1 .$$

The **proper** or **orthochronous Lorentz group** is the subgroup $SO^+(1,3)$
having elements with determinant +1, $\lambda_{11} > 0$.

(ii)   The notation $SO^+(1,3)$ emphasises the sub-Lorentz group that
consists of **Special Orthogonal** transformations of **Minkowski space**
i.e. of R^4 with the Lorentz metric tensor g_o ; special means
orientation-preserving.   In fact, $SO(1,3)$ is an orientation
preserving normal Lie subgroup of the orthogonal group $O(1,3)$.
Recall that $O(3)$, the group of rotations and reflections of R^3,
consists of orthogonal 3×3 real matrices σ satisfying
$\sigma^T I \sigma = I$, that is $\sigma^T = \sigma^{-1}$; and $SO(3)$ is the subgroup
having determinant +1 , i.e. the rotations.

(iii) $SO(1,3)$ has two components but $SO^+(1,3)$ is **connected**.

(iv) $SO^+(1,3)$ consists of the **rotations** of $R^1 \oplus R^3$ that preserve
the orientations of the two factors.

(v) $SO^+(1,3)$ is **non-compact**, it has no finite dimensional
unitary representations and it has infinite irreducible
representations.   This is in contrast with the rotation group
$SO(3)$ for R^3 , which is compact, any representation can be made
unitary and any irreducible representation is finite-dimensional.
Of course, $SO(3)$ is a subgroup of $SO^+(1,3)$ because for each
R^3-rotation $\sigma \in SO(3)$, we obtain a proper Lorentz transformation

$1 \oplus \sigma \in SO^+(1,3)$.

(vi) The <u>universal covering group</u> of $SO^+(1,3)$ is $S\ell(2;C)$ the group of 2×2 complex matrices with determinant +1 .

(vii) $SO(1,3)$ is a <u>closed</u> subgroup of $G\ell(4;\mathbb{R})$.

(viii) As a manifold, $SO(1,3)$ has <u>dimension 6</u> and hence its Lie algebra is 6-dimensional; the same is true for $SO^+(1,3)$.   For comparison: $SO(3)$ has dimension 3, $S\ell(2;C)$ has (real) dimension 6 and $S\ell(2;\mathbb{R})$ has dimension 3.

(ix) The fundamental group of $SO^+(1,3)$ coincides with that of $SO(3)$, it is Z_2 .

1.   LORENTZ STRUCTURES

In order to support relativity theory, or any other plausible geometrical theory of physics (cf. Hawking and Ellis [43] and Misner, Thorne and Wheeler [72]) our manifold needs a Lorentz structure if it is to represent the observable universe.   The peculiar distinction of a Lorentz structure over other metric structures is that it determines a family (or <u>cone</u>) of directions in tangent spaces, these we call <u>timelike</u> directions.   Conversely, if we are given a nowhere-zero vector field

$$v : M \longrightarrow TM$$

on a manifold M , and if M admits a Riemannian structure h then a Lorentz structure h_v is obtained by

$$h_v = h - \frac{2}{h(v,v)} h_\downarrow(v) \otimes h_\downarrow(v)$$

where $h_\downarrow$ is at each $x\in M$ the dualizing isomorphism (cf.[28] for further discussion of $h_\downarrow$ and its inverse $h^\uparrow$)

$$T_x M \longrightarrow (T_x M)^* : a \longmapsto h_x(a,-) = h_\downarrow(a) .$$

This formula for h_v is from Avez [2] and we observe that it is independent of the scale of the field v .   So in fact a one-dimensional distribution (cf. IV §2.2), that is a field of unscaled directions (cf. Geroch [39]), or simply line elements (cf.

Steenrod [96] and Markus [70]), would be sufficient to determine h_v from h . Of course there may well be many choices for h , none of which is particularly distinguished so the field v certainly does not determine a unique Lorentz structure. However, with respect to this h_v , the field v is everywhere <u>timelike</u>. Explicitly,

$$h_v(v,v) \;=\; h(v,v) - \frac{2}{h(v,v)} h(v,v)h(v,v) \;=\; -h(v,v) < 0 \;,$$

since any Riemannian h is positive definite. For any Lorentz structure g , a vector $u \epsilon T_x M$ is called (cf. [28]):

(i)   <u>timelike</u> if $g_x(u,u) < 0$;

(ii) <u>null</u> or <u>lightlike</u> if
      $g_x(u,u) = 0$;

(iii) <u>spacelike</u> if $g_x(u,u) > 0$.

    Now, conversely, we can work from a Lorentz structure g on M to find a field of timelike directions, provided that M also has a Riemannian structure h . For example, from Avez [2] (cf. Geroch [39] p. 79 for a picture of what is happening) consider

$$v_h : M \longrightarrow TM : x \longmapsto \pm v_h(x)$$

where $v_h(x) \epsilon T_x M$ is that unique (up to sign) element u with $h_x(u,u) = 1$ which maximizes $-g_x(u,u)$.

    In summary then we have:

1.1   Theorem

Let (M,h) be a Riemannian manifold, then

(i)   every one-dimensional distribution on M determines a
      Lorentz structure g_v on M with respect to which v is
      timelike;

(ii) every Lorentz structure g on M determines a timelike
      distribution on M .   □

Corollary 1

By Theorem 1 in §2.8, the properties (i) and (ii) will also hold
for any paracompact manifold M because these always admit
Riemannian structures.   ☐

Corollary 2

By the Corollary to Theorem 5 in §2.8, the properties (i) and (ii)
again hold for any manifold M having a connection in its frame
bundle, because that is a (necessary and) sufficient condition for
paracompactness.   ☐

Corollary 3

Since a manifold M is paracompact if and only if each of its
components is second countable (has a countable base for its
topology) properties (i) and (ii) hold for second countable
manifolds.   A sufficient condition for second countability is the
admission of a countable atlas.   (cf. Brickell and Clark [9] p. 43
et seq.)   ☐

Corollary 4

Since a parallelization of an n-manifold is a specification of
n linearly independent vector fields, a parallelizable manifold
has n nowhere-zero vector fields and hence many Lorentz structures.
In Theorem 6 of Ch. IV §2.8 we saw how one of these could be
selected in a unique way.   ☐

We conclude that a nowhere-zero vector field (or at least a
field of directions) and paracompactness are needed to determine a
Lorentz structure.   Conversely, a manifold is paracompact if and
only if it admits a Lorentz structure, by the Corollary to
Theorem 5 in §2.8, so we have

1.2   Theorem

Every spacetime is paracompact.   ☐
(A direct proof is given in Geroch [37] p. 1743-4.)

Corollary 1

Every spacetime is second countable, first countable and separable (cf. III §1.5).   ☐

Corollary 2

Every spacetime admits a Riemannian structure and therefore it has
a metric topology, by Theorems 1 and 3 in §2.8.   ☐

Corollary 3

Every spacetime admits a partition of unity.

Proof

Brickell and Clark [9] p. 51.   ☐

Corollary 4

If U is an open subset of a spacetime M and V is a closed set
contained in U then there exists a smooth real function f on
M taking the value 1 on V and the value 0 on M\U .

Proof

Brickell and Clark [9] p. 51, via a partition of unity. A similar
device is commonly used to smooth out functions given locally on a
coordinate neighbourhood.   ☐

We shall require our spacetimes to be non-compact for good
physical reasons. However, spacetime manifolds are often
constructed as products involving compact manifolds so we note
the following.

1.3   Theorem

Any compact manifold admits a nowhere-zero vector field if and
only if its Euler characteristic is zero.

Proof

Husemoller [52] p. 259, (cf. also Markus [70]).   ☐

Corollary 1

The Euler characteristic of the sphere S^n is $1+(-1)^n$ so only
odd dimensional spheres admit nowhere-zero vector fields. Hence,
for this reason alone, S^4 cannot be provided with a Lorentz
structure, but $S^3 \times S^1$ could be so provided. (Both are of course
inextensible.)   ☐

Corollary 2

Any compact odd dimensional manifold admits a nowhere-zero vector field because its Euler characteristic is zero.

Proof

Steenrod [96] p. 207.   □

1.4   Theorem

If any manifold M_1 admits a nowhere-zero vector field then so does $M_1 \times M_2$ for any manifold M_2.

Proof

Brickell and Clark [9] p. 118.   □

Corollary 1

For any manifold M_o, $R \times M_o$ always admits a nowhere-zero vector field because R does.   □

Corollary 2

The product manifold $R \times S^3$ can be given the structure of a spacetime (it and $R \times R^3$ are the most popular candiates for cosmological models, taking R as the time coordinate).   □

The problem with compact manifolds is seen in the next result.

1.5   Theorem

If g is a Lorentz structure on a compact 4-manifold M then there exist closed curves in M having timelike tangent vectors everywhere.

Proof

Geroch [36], (cf. also Geroch [39] p. 81 for a picture, and Bass and Witten [3]); Tipler [98] gives a stronger result.   □

Corollary

The manifold $S^1 \times S^3$ certainly admits a Lorentz structure, but is unacceptable physically because it is compact and will contain a closed timelike curve; the same is true of $S^1 \times S^1 \times S^1 \times S^1$ for example.   □

From Sard's Theorem (in IV §1.3) we know that the set of
singular points of a smooth map f : M ⟶ R has measure zero image.
But x∈M is such a singular point if at x the 1-form field df
takes the value $\underline{0} \in T_x M^*$.   We get an easy deduction:

1.6   Theorem

Every smooth map f : M ⟶ R determines on a paracompact manifold
M a vector field that is nowhere-zero, except on a set of measure
zero image.

Proof

By hypothesis, df is a section of the cotangent bundle and by
Sard's Theorem it is zero on a set of measure zero image. If this set
contains a point x then in some chart (U,φ) about x we have:

$$df = \partial_i f \, dx^i \quad \text{on} \quad U$$

and $(\partial_i f) = \underline{0} \in R^n$ at x .

Hence we obtain a section $df^\dagger$ of the tangent bundle by attaching
the components $(\partial_i f)$ of df to each frame (∂_i) dual to (dx^i).
That is, about x∈U ,

$$df^\dagger_U = f^i \partial_i \qquad \begin{cases} \text{where} \;\; f^i = \partial_i f \;\; \forall i \\ \text{and} \;\; dx^j(\partial_i) = \delta^j_i \;\; \forall i,j . \end{cases}$$

and we use paracompactness to smooth out these local sections via
a partition of unity.   □

Now we concentrate on non-compact manifolds.   There we can
always provide a Lorentz structure in the presence of paracompactness
because of the following results.

1.7   Theorem

Every non-compact paracompact manifold admits a nowhere-zero vector
field.

Proof

Markus [70] Steenrod [96].   (Cf. Hirsch [49] and also Geroch [39]
p. 80 for an intuitive account of how to push isolated singularities
'out to infinity', i.e. outside any compact set.)   □

Corollary 1

Every non-compact paracompact manifold M admits a Lorentz
structure by Theorem 1.   Then the frame bundle LM is reducible
to OM , the principal fibre bundle of orthonormal frames, with
structure group the Lorentz group SO(1,3) .

Proof

Markus [70] p. 414.   Cf. Chapter IV, Theorem 2.3.2 and note that
the Lorentz group is closed in $G\ell(4;\mathbb{R})$.   □

Corollary 2

On a non-compact paracompact 4-manifold a Lorentz structure can
always be so chosen as to give no closed timelike curves.

Proof

Penrose [75].   □

     We can bring Corollary 1 to bear on some results in the previous
chapter to express the next deductions.

1.8   Theorem

The frame bundle LM over a spacetime (M,g) is reducible to a
principal fibre bundle, with structure group G' a connected Lie
subgroup of the Lorentz group, if and only if the associated bundle
with fibre $G\ell(4;\mathbb{R})/G'$ admits a section

$$\sigma : M \longrightarrow LM/G'$$

Proof

This is Theorem 2 in Ch. IV §2.3, with the added ingredient:

     Lemma   The connected Lie subgroups of the Lorentz group
              are closed in $G\ell(4;\mathbb{R})$.

     Proof   This result is given in Schmidt [83] and discussed
by Friedrich [34].   The following approach was suggested to me by
D.B. Epstein: From Hochschild [50] p. 192, if G' is a connected
Lie subgroup of G then G' is closed in G <=> the closure
of every 1-parameter subgroup is contained in G' .   A 1-parameter
subgroup is either closed or its closure is a torus (cf. [50] p.191).

Now, the Lorentz group is itself closed in $G\ell(4;R)$ so it is
sufficient to show that connected Lie subgroups of the Lorentz
group are closed.   Any torus is contained in a maximal compact
subgroup, this in the case of the Lorentz group is the special
orthogonal group $SO(3)$ which has maximal torus a circle (cf.
Husemoller [52] p. 182).   Hence, every 1-parameter subgroup of
the Lorentz group is either a circle or a closed copy of R , and
it is therefore closed in the Lorentz group and in $G\ell(4;R)$.   □

1.9   Theorem

The structure group $G\ell(4;R)$ of LM for a spacetime   (M,g)   can
be reduced to a subgroup   G'   if and only if there exists a
connection in LM having G' as its holonomy group.   Therefore,
LM admits a connection having the Lorentz group as its holonomy
group.

Proof

This is a consequence of Theorem 5 in Ch. IV §2.6, because of
Corollary 1 to Theorem 7 in our present section.   □

2.   ORIENTABILITY

We recall that a frame   (x,b)   at a point   x   in a manifold   M   is
an ordered basis b_1 , b_2 , ..., b_n for $T_x M$.   The set of such
frames at x is the fibre $\overleftarrow{\Pi}_L(x)$ of the frame bundle LM (cf.
IV §2.3).   Any two such frames   (x,b), (x,b')   are related through
a real non-singular matrix $[h^i_j]$ by

$$b'_j = h^i_j b_i .$$

Fundamental to the notion of orientability is the equivalence
relation $\sim x$ on $\overleftarrow{\Pi}_L(x)$ given by :

$$(x,b') \sim x (x,b) \iff b'_j = h^i_j b_i \text{ with } \det[h^i_j] > 0 .$$

It partitions $\overleftarrow{\Pi}_L(x)$ into two equivalence classes and each class
is called an orientation of $T_x M$.   A chart   (U,φ)   about any   x∈M
determines a frame $(\partial_1)_x = (\partial_1, \partial_2, ..., \partial_n)_x$.   If also   x   lies
in the domain of another chart   (U',φ')   giving a frame $(\partial'_1)_x$ then
the non-singular matrix function relating them on U∩U' is the
Jacobian matrix;   if it has positive determinant everywhere on U∩U'

then we say that the two orientations, $[(\partial_i)_x]_{\sim x}$ and $[(\partial_i')_x]_{\sim x'}$ agree for $x \in U \cap U'$. In that case we obtain an orientation of $T_y M$ for all $y \in U \cap U'$; that is, an orientation on $U \cap U'$.

A manifold M is orientable if it admits an atlas giving an orientation on the whole of M ; with such a choice of atlas M is called an oriented manifold. Like dimension, orientability is invariant under diffeomorphisms.

Evidently, if M admits a (continuous) section of LM then it is orientable, just as if it had an atlas consisting of one global chart (as does $G\ell(n;\mathbb{R})$ for example) so we have:

2.1   Theorem

Every parallelizable manifold is orientable.   □

Corollary

The following are parallelizable and so orientable:

(i)   every open submanifold of $\mathbb{R}^n$;

(ii)   every Lie group ;

(iii) the spheres S^1, S^3 and S^7 ;
       (cf. below)

S^1

(iv)   a product of two or more spheres if at least one has odd dimension (cf. Hirsch [49] p. 98, and Examples below).

(v)   every product of parallelizable manifolds.   □

The situation is also straightforward if M admits an atlas consisting of just two charts:

2.2   Theorem

If $\{(U,\phi), (U',\phi')\}$ is an atlas for M and $U \cap U'$ is connected then M is orientable.

Proof

The Jacobian determinant function cannot change sign on $U \cap U'$. We can therefore find a two chart atlas making this determinant positive, if necessary by changing the sign of one of the coordinates determined by (U',ϕ').   □

Corollary

All spheres are orientable because they admit the two-chart atlas
of stereographic projections (cf. Porteous [78] p. 169).   ☐

    There is an equivalent specification of orientability as
follows (cf. Singer and Thorpe [90] p. 128).   Suppose that M
has an atlas $\{(U_\alpha,\phi_\alpha)|\alpha\epsilon A\}$.   Then M is <u>orientable</u> if it admits
a subatlas indexed by A' such that whenever (x^i) , (y^i) are
coordinates with respect to ϕ_α , ϕ_β on $U_\alpha \cap U_\beta$ for $\alpha,\beta\epsilon A'$ then
the real function f on $U_\alpha \cap U_\beta$ defined by

$$(dx^1 \wedge \ldots \wedge dx^n)_x = f(x)(dy^1 \wedge \ldots \wedge dy^n)_x$$

is everywhere positive.   We observe that this is an equation
between n-forms on $U_\alpha \cap U_\beta$ (cf. IV §2.7).   In fact, orientability
is completely characterized by the existence of a <u>nowhere-zero</u>
<u>n-form on</u> M , also called a <u>volume element on</u> M (cf. IV §2.7):

2.3   <u>Theorem</u>

Let M be an n-manifold;   then

(i)   M is orientable if it admits a volume element,

(ii)   M has a volume element if it is orientable and paracompact.

Proof

Singer and Thorpe [90] p. 129-134.   ☐

    The great advantage of this result is that it allows non-
orientability to be revealed by showing that every (continuous)
n-form is zero somewhere, usually a much simpler process than
showing the non-existence of an oriented atlas.   There is a
<u>standard volume element</u> determined by the metric tensor field on
any Riemannian or pseudo-Riemannian oriented manifold, and its
relation to the Hodge star operator (cf. IV §2.7) is indicated
next.

2.4   <u>Theorem</u>

Given an oriented n-manifold M and on it a metric tensor field g
with eigenvalues

$$\underbrace{(-1, \ \ldots, \ -1,}_{b} \ \underbrace{1, \ \ldots, \ 1)}_{n-b}$$

then the standard volume element is given in local coordinates by

$$v \ = \ \left|\det g_{ij}\right|^{\frac{1}{2}} dx^1 \wedge \ldots \wedge dx^n$$

where

$$g \ = \ g_{ij} \ dx^i \otimes dx^j \ .$$

The following results hold:

(i) $*1 = v$,   (here of course 1 is a constant element

            of $\Lambda T_o M = R$) ;

(ii) $**1 = *v = (-1)^b$;

(iii) for any r-form w, $**w = (-1)^{r(n-r)+b} \, w$;

(iv)   g determines a unique dual $g^\dagger \in T^2 M$ and for all 1-forms
       λ, μ it follows that

$$*(\lambda \wedge *\mu) \ = \ g^\dagger (\lambda,\mu) **1 \ = \ (-1)^b \ g^\dagger (\lambda,\mu) \ .$$

<u>Proof</u>

(i)   By non-degeneracy of g it follows that in any coordinates
$\det g_{ij} \neq 0$,   hence v is a volume form.   Then $*1 = v$ is given
by definition.

(ii)   Since * is linear, and we can work locally

$$**1 = *v = \left|\det g_{ij}\right|^{\frac{1}{2}} *(dx^1 \wedge \ldots \wedge dx^n)$$

$$= \left|\det g_{ij}\right|^{\frac{1}{2}} \left|\det g_{ij}\right|^{\frac{1}{2}} (\det(g_{ij})^{-1}) \quad (\text{cf. (iii)})$$

$$= \left|\det g_{ij}\right| (\det(g_{ij})^{-1}) \ = \ (-1)^b \ .$$

(iii) The Hodge operator has the following action on any r-form
$w = w_{(i_1 \ldots i_r)} dx^{i_1} \wedge \ldots \wedge dx^{i_r}$ (cf. Flanders [33])
$*w = w_{(j_1 \ldots j_{n-r})} dx^{j_1} \wedge \ldots \wedge dx^{j_{n-r}}$

where

$$w_{j_1 \ldots j_{n-r}} = \left|\det g_{ij}\right|^{\frac{1}{2}} w^{(k_1 \ldots k_r)} \, \text{sgn}(k_1, \ldots, k_r, j_1, \ldots, j_{n-r})$$

and

$$w^{k_1 \ldots k_r} = g^{k_1 i_1} \ldots g^{k_r i_r} w_{i_1 \ldots i_r}$$

with $(g_{ij})^{-1} = (g^{ij})$.

Now, by the product rule for determinants

$$(\det g_{ij})(\det g^{ij}) = 1 ,$$

also

$$\det g^{ij} = \sum_{(i_1 \ldots i_n)} \text{sgn}(i_1, \ldots, i_n) \, g^{1 i_1} \ldots g^{n i_n} .$$

The result follows, after some manipulation.

(iv) The construction of $g^\dagger$ is discussed in Dodson and Poston [28]. Its components are $(g^{ij}) = (g_{ij})^{-1}$. Given $\lambda = \lambda_i dx^i$, $\mu = \mu_j dx^j$ we have

$$g^\dagger(\lambda, \mu) = g^{ij} \lambda_i \mu_j .$$

From (iii)

$$*\mu = |\det g_{ij}|^{\frac{1}{2}} \sum_m g^{mj} \mu_j \, \text{sgn}(m, i_1, \ldots, i_{n-1}) dx^{i_1} \wedge \ldots \wedge dx^{i_{n-1}}$$

and so

$$\lambda \wedge *\mu = |\det g_{ij}|^{\frac{1}{2}} g^{ij} \lambda_i \mu_j \, dx^1 \wedge \ldots \wedge dx^n$$

hence the result.   □

Corollary 1

In the case $n = 3$ and $M = \mathbb{R}^3$ with the standard metric structure (b = 0), the cross product $\times$ for geometrical vectors $\lambda, \mu \in \Lambda \mathbb{R}^3 = \mathbb{R}^3$ is related to $\wedge$ by :

$$*(\lambda \wedge \mu) = \lambda \times \mu .   \square$$

Corollary 2

In the familiar case of M being a 2-dimensional oriented submanifold of $\mathbb{R}^3$ we can see the role of the standard volume element on (M, g), where g is the Riemannian structure induced on M by the Euclidean geometry of $\mathbb{R}^3$. Given a chart (U, ϕ) on M with coordinates (x^1, x^2), the usual area of the set U

is

$$\text{Area (U)} \quad = \quad \int\limits_{(x^i)\,\epsilon\,\phi U} \int \; |\det g_{ij}|^{\frac{1}{2}} \; dx^1 dx^2 \; .$$

This expression generalises to subsets of $I\!R^n$ referred to arbitrary coordinates.   Since the components of g change under a local coordinate change $(x^i) \longmapsto (\bar{x}^i)$ according to

$$g_{ij} dx^i \otimes dx^j \; = \; \bar{g}_{ij} d\bar{x}^i \otimes d\bar{x}^j$$

we can also see why the Jacobian determinant arises in the elementary treatments of multiple integrals.   □

Remark

Applications of the volume form in spacetime geometry are given in [72] and [101].

Examples

(i)   The following are <u>orientable</u> manifolds:

(a)   TM for any manifold M   (cf. Greub, Halperin and Vanstone [40] p. 129) ;

(b)   the torus $S^1 \times S^1$;

(c)   every 1-manifold ;

(d)   the product of orientable manifolds ;

(e)   real projective n-space $I\!RP^n$ for all odd n .

(ii)   The following are <u>non-orientable</u> manifolds:

(a)   the Mobius strip ;

(b)   the Klein bottle ;

(c)   the product of any manifold with a non-orientable manifold ;

(d)   real projective space $I\!RP^n$ for any even n⩾2 .

For more discussion see Singer and Thorpe [90] and Spivak [94] .

The next result shows how to construct an orientable covering manifold over a non-orientable manifold.

2.5   Theorem

If M is a connected non-orientable manifold then there exists a connected orientable manifold $\tilde{M}$ which is a twofold covering of M.

Proof

(After Schwartz [87] p. 14.)

Let $\{(U_\alpha, \phi_\alpha) | \alpha \epsilon A\}$ be an atlas for M and define

$$S = \{(x, U_\alpha) | \alpha \epsilon A, x \epsilon U_\alpha\} .$$

We take two copies of S, S_+ and S_- say.   Denote by $|J_{\alpha\beta}|_x$ the determinant of the Jacobian matrix of the change of coordinates from chart α to chart β at $x \epsilon U_\alpha \cap U_\beta$.   We introduce a relation $\sim$ on $S_+ \cup S_-$ by :

$$(x, U_\alpha)_+ \quad \sim \quad (y, U_\beta)_+ \quad \Longleftrightarrow \quad x=y \text{ and } |J_{\alpha\beta}|_x > 0 ;$$

$$(x, U_\alpha)_+ \quad \sim \quad (y, U_\beta)_- \quad \Longleftrightarrow \quad x=y \text{ and } |J_{\alpha\beta}|_x < 0 ;$$

$$(x, U_\alpha)_- \quad \sim \quad (y, U_\beta)_- \quad \Longleftrightarrow \quad x=y \text{ and } |J_{\alpha\beta}|_x > 0 .$$

This is an equivalence relation and its equivalence classes constitute the elements of $\tilde{M}$.   A topology is provided by taking, for example,

$$\{[(y, U_\alpha)_+]_\sim \mid y \epsilon U_\alpha\}$$

as a neighbourhood of $[(x, U_\alpha)_+]_\sim \epsilon \tilde{M}$.

A manifold structure is provided by :

(i)   about $[(x, U_\alpha)_+]_\sim$ use the coordinates $(x^1, x^2, \ldots, x^n)$ of ϕ_α

(ii)   about $[(x, U_\alpha)_-]_\sim$ use the coordinates $(x^1, x^2, \ldots, x^{n-1}, -x^n)$.

It follows that $\tilde{M}$ is orientable, and therefore connected for essentially the same argument as in Theorem 4.   Also,

$$f : \tilde{M} \longrightarrow M : [(x, U_\alpha)_\pm]_\sim \longrightarrow x$$

is a two to one, differentiable, covering map.   □

      We have seen that on a paracompact manifold M the existence
of a Lorentz structure is equivalent to the existence of a
1-dimensional distribution on M .   Such a distribution is called
an orientable line element field if it is generated by a continuous
tangent vector field; in that case there is available a continuous
assignation of a 'forward' direction at each point.   A necessary
and sufficient condition for a line element field to be orientable
is given by the following result of Markus [70] .

2.6   Theorem

A continuous line element field L on a paracompact, separable
manifold M is orientable if and only if the orientation of L
is preserved around each of a set of generators of $\pi_1(M)$, the
first homotopy group (cf. III §1.7).   If L is not orientable
then M has a twofold covering $\tilde{M}$ and L can be lifted to an
orientable line element field $\tilde{L}$ on $\tilde{M}$.   If $\pi_1(M)$ contains
no proper subgroup of index two, then every continuous line element
field on M is orientable.   □

(The index of a subgroup H of a group G is the number of (right
or left) cosets of H in G .)
Markus [70] also pointed out that if $\pi_1(M)$ contains no proper
subgroups of index two then every line element field L that is
continuous except for isolated singularities is orientable if n⩾3.
For in this case $\pi_1(M)$ is isomorphic to the fundamental group of
the open submanifold of M on which L is defined and continuous.
Therefore L is always orientable on a simply-connected manifold
(which is then its own universal covering manifold), or if n⩾3 ,
in the locality of an isolated singularity.

      We shall say that a spacetime (M,g) is time-orientable
(also called isochronous) if the line element field determined by
g (Theorem 1 in §1) is orientable.   Following from the results
of Markus we deduce:

2.7   Theorem

Every spacetime $(\tilde{M},\tilde{g})$ admits a covering spacetime (M,g) which
is time-orientable.   □

Remarks

(Cf. Geroch [36], where $(\tilde{M},\tilde{g})$ is called the Lorentz covering manifold.)

(i)   We actually construct $\tilde{M}$ from an equivalence relation on the set, for any fixed $p \in M$,

$$S_p = \{(x,c) \,|\, x \in M, \text{ c is a continuous curve from x to p}\}.$$

Put:

$$(x,c) \sim (y,c') \iff \begin{cases} x=y \text{ and the time direction is} \\ \text{preserved on the path from p to x} \\ \text{to p using c and c'} . \end{cases}$$

We assume of course that g has been used to provide a line element field on M and that a choice of forward direction for time has been made, though in general a continuous such choice may be impossible. The factoring $\tilde{M} = S_p/\!\sim$ then effectively eliminates these discontinuities. Since M is by hypothesis connected it is arc connected and so the choice of base point p is immaterial, up to diffeomorphism. Evidently, in a physical sense, $\tilde{M}$ represents the same universe as M .

(ii) If (M,g) is time-orientable then $\tilde{M}$ is diffeomorphic to M , because the curves in S_p are factored out by $\sim$. The differentiable structure is provided on $\tilde{M}$ in a similar way to that employed in Theorem 5.

(iii) Observe that g induces a line element field L on M and, by Theorem 6, L lifts to $\tilde{L}$ on $\tilde{M}$; also $\tilde{M}$ is paracompact and hence by Theorem 1 in §1 $\tilde{L}$ induces many Lorentz structures on $\tilde{M}$. However, we can use g directly to determine a unique Lorentz structure $\tilde{g}$ on $\tilde{M}$ because our construction of $\tilde{M}$ uses copies of charts only trivially different from those on M .

(iv) The construction of $(\tilde{M},\tilde{g})$ ensures that it is the smallest time-orientable covering manifold of (M,g) , since every other candidate also covers $(\tilde{M},\tilde{g})$.

At each point x in a spacetime (M,g) the Lorentz structure allows a splitting of $T_x M$ into timelike, null and spacelike

directions (cf. §1 above).   Hence at each x∈M we can choose a
spacelike <u>triad</u>, an ordered triple of linearly independent spacelike
vectors, because in the presence of g a timelike distribution
determines an orthogonal spacelike one.   Next we can enquire if
such a choice can be made continuously over M in such a way as
to preserve orientation of the spacelike triads round all closed
curves, if so then (M,g) is <u>space-orientable.</u>   It turns out
(cf. Geroch [39]) that time- and space-orientability are independent
of one another, and also independent of the physical notion of
<u>charge-orientability</u> arising from the transport of elementary charges
around closed curves.   There are theoretical reasons for believing
that physical interactions are <u>charge-parity-time-invariant</u>; this
means that either all of charge-, space- and time-orientation
reverse round a closed curve or none of them does.   The belief is
based on the Charge Parity Time or CPT theorem (cf. Geroch [39]
p. 84 and references there) and encouraged by experimental evidence.
It follows that if we wish to impose time-orientability on (M,g)
then to preserve CPT-invariance we must also have space-orientability,
and charge orientability; in that case M is orientable.
The mathematical situation is revealed by the following result:

2.8   Theorem

Let M be a connected paracompact n-manifold having a distribution
L of dimension p and a distribution S of dimension (n-p)
such that TM is spanned by L ⊕ S .   Then any two of the
following statements implies the third :

(i)   M is orientable ,

(ii)   L is orientable ,

(iii) S is orientable .

Proof

Either using algebraic topology or by comparing volume forms for
orientable distributions (cf. Theorem 3); both methods are
outlined by Whiston [104].   ☐

Corollary

For a spacetime   (M,g)   any two of the following statements implies
the third:

(i)   M is orientable ,

(ii)   M is time-orientable ,

(iii) M is space-orientable.   □

It is normal to assume that spacetime is time-orientable and
then at each point x∈M there is determined a division of all
non-spacelike vectors into two classes: forward-pointing and
past-pointing (we allow only the zero vector to be in both).   Then
curves leaving from x with tangent vectors in these classes are
called forward-going or past-going, respectively.   Hawking and
Ellis, [43] chapter 6, provides a thorough study of this causal
structure by investigating the following possible global constraints
on a time-orientable spacetime (M,g) :

(i)   chronology condition : absence of closed timelike curves ;

(ii)   causality condition : absence of closed non-spacelike curves ;

(iii) stable causality condition : g lies in an open set in the
C^0 open topology on metric tensor fields and no metric in this
open set admits closed timelike curves.

Now, in general relativity the Lorentz structure g is
determined by the disposition of matter.   It would therefore be
comforting to think that stability in the sense of constraint (iii)
is present in such a way as to allow for possible quantum
fluctuations in g , when a suitable theory is agreed for their
incorporation.   For this reason we note the interesting necessary
and sufficient condition established by Hawking and Ellis [43]
p. 198-201.

2.9   Theorem

A time-orientable spacetime (M,g) satisfies the stable causality
condition if and only if there is a real function f on M whose
gradient (i.e. df) is everywhere timelike.   □

Remarks

(i)   The intuitive view of stable causality is that a slight
expansion of the cone of forward going curves does not result in
any of them returning to an earlier point.

(ii)   The function f can be viewed as giving universal time.
Hence, if f exists, the forward cone through x∈M (in the
direction of df there) meets a surface of constant f through x
only at x.   The converse, establishing such an f in the presence
of stable causality depends on constructing a measure on M , via
a partition of unity, so f is not uniquely fixed.

(iii) We know that since M has the Lorentz structure g there is
a timelike distribution L on M (by Theorem 1 in §1), though it
is not uniquely determined.   Since (M,g) is time-orientable, L
is orientable and so determined by a nowhere-zero timelike vector
field w on M .   Now in the presence of g, w determines a
unique nowhere-zero timelike 1-form $g_\downarrow(w)$ on M (cf. Dodson
and Poston [28] p. 263).   From Theorem 9 we can deduce that (M,g)
satisfies the stable causality condition if $g_\downarrow(w)$ can be chosen
to satisfy

$$g_\downarrow(w) \quad = \quad df$$

for some real function f on M , that is if $g_\downarrow(w)$ is an _exact_
1-form (cf. IV §2.7).   For this to happen we certainly want $g_\downarrow(w)$
to be _closed_, that is

$$dg_\downarrow(w) \quad = \quad 0 .$$

It is known (cf. Bishop and Goldberg [7] p. 175) that if $g_\downarrow(w)$ is
closed then _locally_ it is exact, so a suitable function exists on
some subset of the domain for each chart from the atlas for M .
In fact this is not saying much because we know that there are
always normal coordinate neighbourhoods about each point (cf. [28]
p. 356).

(iv) For each t∈R the surfaces $f^+(t)$ can be viewed as surfaces
of simultaneity in spacetime, but this may not be very useful unless
some particular f is distinguished by further physical information.
If _all_ of the surfaces $f^+(t)$, t∈R , are _compact_ then they are

all diffeomorphic by following the integral curves of $g^{\uparrow}(df)$; this need not be true if some of them are non-compact.

3.   PARALLELIZABILITY

The main result that we shall be discussing here is due to Geroch [37]:

3.1   Theorem

A space- and time-oriented spacetime (M,g) is parallelizable if and only if it admits a spinor structure.   □

In fact, this does not extend to orientable compact manifolds carrying a Lorentz structure.   Examples of compact non-parallelizable spinor spacetimes have been given by Whiston [105], who also provided the following necessary and sufficient condition from algebraic topology.

3.2   Theorem

A compact spinor spacetime is parallelizable if and only if its Pontryagin number (cf. e.g. Spanier [92]) is trivial.   □

We need a definition (cf. Geroch [37] and Whiston [105]):

A spacetime (M,g) is said to admit a spinor structure or is a spinor spacetime if we have the following:

    (a)   a principal fibre bundle $\widetilde{OM}$ over M , with structure group $S\ell(2;C)$;

    (b)   a two-to-one principal fibre bundle morphism

$$\sigma : \widetilde{OM} \longrightarrow O^{+}M ,$$

where $O^{+}M$ is the principal fibre bundle of oriented orthonormal frames with structure group $SO^{+}(1,3)$, the proper Lorentz group. (Cf. Porteous [78] p. 161 ; as a matrix group, $SO^{+}(1,3)$ is that connected subgroup of the Lorentz group SO(1,3) having $\lambda_{11} > 0$, determinant +1 .)

    We do not propose more than a superficial coverage of the background to Theorem 1 because it, and related results, juxtapose two bodies of theory of very different types that we do not have space to develop here.   On the one hand there is the mathematical theory of characteristic classes within algebraic topology, and on

the other hand there is the physical motivation for spinors
within quantum field theory.   A feel for the mathematical aspects
of the situation is offered through the following notes and
references.

(i)   The group $S\ell(2;C)$ consists of unimodular 2×2 complex
matrices and it is the universal covering group of the proper
Lorentz group.   (Cf. Gel'fand, Minlos and Shapiro [35].)   Physicists
view $S\ell(2;C)$ as the transformation group of spinors, i.e. of
certain elements from a 2-dimensional complex vector space (cf.
Penrose [75], Pirani [76], Crumeyrolle [19] and Cartan [10].)

(ii)   One necessary and sufficient condition for a time- and space-
oriented   (M,g)   to admit a spinor structure is the following
isomorphism of fundamental groups:

$$\pi_1(O^+M) \equiv \pi_1(M) \oplus \pi_1(SO(1,3)).   \text{(N.B. } \pi_1(SO(1,3)) = Z_2)$$

Another such condition is that each of its covering manifolds admits
a spinor structure.   From the results in §2 we see that, unlike
non-orientability, failure to admit a spinor structure cannot be
remedied by passing to a covering space.   (Cf. Geroch [37].)

(iii) The hard part of Theorem 1 is in constructing a parallelization
from a spinor structure.   Just why it goes through cannot be better
put than in Geroch's own words ([37] p. 1743):   "The theorem depends
critically on the vanishing of homotopy groups of the spinor group
$S\ell(2;C)$.   The first homotopy group vanishes essentially because a
spinor structure is defined by the property that its fiber is the
universal covering space of the fiber of the bundle of frames.
(Taking the universal covering group automatically annihilates the
first homotopy group.)   The second homotopy group vanishes for all
the spin groups (in fact, for all Lie groups).   The third homotopy
group fails to vanish, but at this point we are sufficiently close
to the dimension of the manifold that the obstruction to extending
a cross section can be made to vanish.   Thus the dimension of the
manifold enters in an essential way.   In fact, the theorem is
true in four dimensions, uninteresting in lower dimensions (in
this case, every orientable[3]manifold is parallelizable), and false
in higher dimensions."   (Cf. also Lee [63] p. 425 footnote † and

Clarke [14].)

(iv)   In a sequel to his paper quoted above, Geroch [38] found
other tests for the admissibility of spinor structures.   On the
geometrical side, and particularly interesting for the physics of
general relativity, is the result that "a certain threshold of
curvature must be exceeded before there can be even the possibility
of a spacetime having no spinor structure."   This is surprising in
view of the known topological character of conditions for
parallelizability.   Also, Geroch gave a topological condition for
a time- and space-oriented spacetime (M,g) to be parallelizable.
It is sufficient that M is topologically the product of R with
a spacelike 3-surface; hence if M is constructible from initial
data on a Cauchy surface (cf. Hawking and Ellis [43] p. 201 et seq.)
then it is parallelizable.   At work here is the trivial
parallelizability of R , and the parallelizability of any
orientable 3-manifold (cf. Steenrod [96] p. 203 and 221).   In view
of Theorem 9 in the previous section we have:

3.3   Theorem

A space- and time-orientable spacetime (M,g) is parallelizable
if it satisfies the stable causality condition and if a universal
time function f : M $\longrightarrow$ R so determined has homeomorphims

$$\overleftarrow{f}(t) \;\approx\; \overleftarrow{f}(t') \qquad \forall t,t' \in R \;. \qquad\qquad \Box$$

Corollary

The required homeomorphisms are necessarily present (and in fact
diffeomorphisms) if $(\forall t \in R)$ $\overleftarrow{f}(t)$ is compact. $\Box$

4.   PRODUCT SPACETIMES

We make a few points about a simple class of spacetimes and mention
some restrictions on the evolution of the topology of spacelike
sections.

    The simplest candidate for a spacetime (M,g) , to model part
or all of the universe, is a product of Riemannian manifolds
(S,g_σ), (R,g_τ) in the form

$$M = \mathbb{R} \times S , \quad g = (-g_\tau) \otimes g_\sigma .$$

Here, S is some 3-manifold representing the space of positions at a given instant of time. This is a simple model because both the topology and the geometry of the space of positions remains unaltered throughout all time, representing a static universe.

The next level of complexity is to keep the product topology, $M = \mathbb{R} \times S$, but to relax the product geometry by taking

$$g \neq (-g_\tau) \otimes g_\sigma .$$

Typically, the geometry of the spacelike sections $\{t\} \times S$ are made to depend on t and the simplest way is to introduce a conformal factor (cf. Hawking and Ellis [43]) to give

$$g = \rho^2 . (-g_\tau) \otimes g_\sigma$$

where ρ is some positive function (of time $t \in \mathbb{R}$ only, here) which intuitively serves to measure the 'radius' of space $\{t\} \times S$ at each instant $t \in \mathbb{R}$. Thus, to model the observed Hubble redshift of stellar spectra we represent a universal expansion of space by choosing ρ to be a strictly increasing function of time. Particular examples of spacetimes of these types and others conformal to products are discussed in Hawking and Ellis [43] Ch. 5, (cf. also Misner, Thorne and Wheeler [72], Sachs and Wu [82] and Clarke [17]).

From our previous results we collect the following.

4.1   Theorem

Let a spacetime (M,g) have M diffeomorphic to a product of $\mathbb{R}$ (representing time) with a 3-manifold S (representing space). Then we can deduce:

(i)   If S is orientable then S is parallelizable so LS is trivial; also M is orientable, time-orientable and space-orientable and parallelizable hence LM is trivial.

(ii)   If S is not orientable then S is not parallelizable and so LS is non-trivial, M is not orientable, not space-orientable, not parallelizable and LM is non-trivial.

(iii) (M,g) satisfies the stable causality condition.

Proof

(i) Since R is orientable, if S is orientable then M is orientable; and M is time-orientable, therefore by Theorem 8 in the previous section, M is space-orientable. Any orientable 3-manifold is parallelizable (cf. Steenrod [96] p. 203, 221). Finally, LS is trivial if S is parallelizable by Theorem 3 Ch. IV §2.3.

(ii) If S is parallelizable then so is M because R is parallelizable. Being parallelizable, M is orientable; but since R is orientable, M is time-orientable and therefore also space-orientable. Again, parallelizability of M ensures triviality of LM .

    If S is not orientable then M is neither orientable nor space-orientable. Since any parallelizable manifold is orientable, S and M are not parallelizable and hence LS and LM are non-trivial.

(iii) Since $M = R \times S$ is time-orientable, stable causality is assured by Theorem 9 in §2 if we can find a universal time function. Such a function is given by projection onto the R component. □

    The subsets $\{t\} \times S$ for $t \epsilon R$ are   prototypes of Cauchy surfaces (cf. Hawking and Ellis [43] p. 205). For given $t_0 \epsilon R$ then any $(t',x') \epsilon R \times S$ lies on the timelike curve

$$c : R \longrightarrow M : t \longmapsto (t,x')$$

through $(t_0,x') \epsilon \{t_0\} \times S,$

and each inextensible non-spacelike curve through (t',x') meets $\{t\} \times S$ just once.

    Evidently, if $M \equiv R \times S$ and S is orientable then M has a trivial structure for its bundles because by parallelizability we deduce that

$$LM = R \times S \times G\ell(4;R) .$$

Furthermore, by Corollary 1 of Theorem 1 in §1, LM is reducible to the orthonormal frame bundle OM , with structure group the

Lorentz group.

Typical candidates for the Riemannian 3-manifold S in a product spacetime with $M \equiv \mathbb{R} \times S$ are $\mathbb{R}^3$, S^3 and possibly also products like $S^2 \times \mathbb{R}$, $S^1 \times S^1 \times \mathbb{R}$. Through a study of the homotopy groups associated with the classical groups (orthonormal, unitary and symplectic) Husemoller [52] Ch. 7 gives a classification of all vector bundles over spheres of dimension up to 4. Topologically:

$$S^1 \simeq SO(2), \ \mathbb{R}P^3 \simeq SO(3) \simeq S^3/Z_2$$
$$S^3 \simeq SU(2), \ SO(4) \simeq S^3 \times SO(3)$$

and we have the fundamental groups :

$$\pi_1(SO(2)) = Z, \ \pi_1(SO(3)) = Z_2 = \pi_1(SO(4)),$$

also, of course,

$$\pi_1(S^1 \times S^1) = Z \oplus Z, \ \pi_1(S^3) = \pi_1(\mathbb{R}^n) = \pi_1(S^2) = 0.$$

Husemoller's classification is through a consideration of bundles having the classical groups as fibres. Whiston [105] has also studied the Lorentz group and the associated spinor groups; here we note his observation that the inclusion

$$SO(3) \hookrightarrow SO^+(1,3)$$

which we encountered earlier is actually a homotopy equivalence and so induces an isomorphism

$$\pi_1(SO^+(1,3)) \equiv \pi_1(SO(3)) = Z_2.$$

In fact, much material is available generally on the classification of the product spacetimes of likely interest, because of the work that has been done for 3-manifolds. Even with relatively simple topologies there is of course considerable scope left for supplying interesting geometries through the Lorentz structure. This is illustrated by the analyses in Hawking and Ellis [43] of spacetimes having Cauchy surfaces.

It is natural to enquire whether a change in the topology of spacelike regions can occur with time in a physically reasonable, general spacetime. In fact, the question is only meaningful in those spacetimes that admit a splitting into space and time in the locality of interest. One technical construction that is used in

these situations is a <u>partial</u> Cauchy surface.   We do not propose
to become involved with the details, for they are thoroughly
covered by Hawking and Ellis [43] and similar methods are used in
the references we mention below.   We shall merely describe
informally the main results that restrict unusual evolution of the
topology of spacelike regions.   Logically this material fits here
because it pertains to spacetimes that are natural generalisations
of global products.   Additionally, we shall see that under
reasonable conditions a change in
the topology of a spacelike region
will result in some curve becoming
inextensible while still in some
sense finite; we conclude that it
meets a <u>singularity</u>.   Singularities
are the subject of the next section.

     Geroch [36] showed that a change
of topology of a compact spacelike
region with time in a compact spacetime
is impossible if we do not allow closed
timelike curves.   Also, Lee [61] established that if a region of
space can be enclosed by a spacelike C^2-embedding of S^2 having
one congruence (ingoing, say) of null vectors orthogonal to it
converging (a <u>uniformly convex sphere</u>), then that region is compact
and simply connected; moreover, modulo the Poincaré conjecture,
it is homeomorphic to the unit 3-ball.   Conversely, if some non-
trivial topology of space could be enclosed by a uniformly convex
sphere then one of the ingoing null geodesics orthogonal to the
enclosing sphere would be expected to be inextensible after a time
of the same order as the diameter of this sphere.   Physically,
an ingoing photon would cease to exist after a finite time.
Precisely this happens in the case of the uniformly convex spheres
enclosing a Schwarzschild-Kruskal black hole.   These spheres have
non-compact interior and hence a non-trivial topology.
Calculations (cf. Misner, Thorne and Wheeler [72]) show that for
a black hole of solar mass (about 10^{30} kg) the radius of the sphere
coinciding with the 'event horizon' is a few kilometres, so an

ingoing photon passing through the
event horizon meets the central
singularity after a flight time of
about ten microseconds.

   Tipler [98] extended Geroch's
result by showing that the causality
condition was not necessary if
gravity is assumed to be everywhere
attractive and subject to Einstein's
equation.   Specifically, if this
latter assumption is true on a compact
subset B bounded by the disjoint union of two compact 3-manifolds
S_1 and S_2 then $S_1 \equiv S_2$ and $B \simeq S_1 \times [0,1]$.   Moreover, if a
topology change does occur in a 4-dimensional region B ,   then
B is not compact and it contains a singularity if B is 'finite'.
Lee [62] showed that closed and bounded regions of space cannot
change topology with time in a spacetime that is timelike and null
geodesically complete with gravity everywhere attractive and no
closed timelike curves.

   Everyday experience suggests to us that in small enough
regions we can decompose the observable universe into a succession
of spacelike slices parametrized by time, and the spacelike sections
have persisting, trivial topologies.   More sophisticated experiments
interpreted through very reliable theories concerning the behaviour
of matter, have convinced cosmologists that the observable universe
does indeed contain singularities in the classical sense.   We end
this section with notes on the three principal cases.

(i)   There is overwhelming evidence to support the view that our
universe had a 'beginning' some 10^{10} years ago.   In particular,
certain photons (collected in detectors of the isotropic background
black body radiation) yield examples of finite null geodesics that
are past-inextensible (beyond the primeval Big Bang).

(ii)   The final state of some stars is expected to be a black hole,
achieved through a collapse of molecular, atomic and nuclear matter
under gravitational forces.   Astronomers are convined that no
alternative explanation remains for the superdense dark companion

of Cygnus Xl.

(iii) The turbulent conditions during the early stages of the universe could have generated black holes over a range of masses. Recently, Hawking has shown that certain quantum theoretical effects are particularly important for small black holes : they 'evaporate' more quickly than large ones.   The process is one of losing energy by creating particles in the strong fields surrounding their event horizons.   So eventually black holes disappear completely because the rate of energy loss continues to increase, but the process is very slow indeed for large black holes.   Now, we believe that the Big Bang was about 10^{10} years ago and this turns out to be the expected liftime of a black hole of mass about 10^{12} kg;   it has an event horizon commensurate with nuclear radii.   These 'mini black holes' have not yet been detected, but the theoretical basis for their existence continues to be strengthened.

5.   SINGULARITIES

We found in the previous section that rather mild topological anomolies could lead to incompleteness and singularities.   Hawking and Ellis [43] show quite generally that, in any physically reasonable spacetime, timelike or null geodesic incompleteness is very likely. In relativity theory, timelike and null geodesics are viewed as potential trajectories for 'free' material particles and photons, respectively. By 'free' is meant an absence of external forces. Experimentally, a freely falling satellite seems to follow very precisely a timelike geodesic in a Schwarzschild spacetime used to model a neighbourhood of the Earth. However, a rocket used to place the satellit in orbit

is certainly not falling freely and it follows a curve that is
timelike but not necessarily a geodesic.   In fact, whereas timelike
or null geodesic incompleteness would certainly be sufficient for a
spacetime to admit singularities in a physical sense, it may not be
necessary.   For there is an example of a timelike and null
geodesically complete spacetime that contains incomplete timelike
curves which could represent the trajectory of a rocket having
bounded acceleration.   This is in marked contrast to the situation
for Riemannian manifolds, where geodesic completeness is actually
equivalent to metric completeness (cf. Theorem 4 in Ch. IV §2.8).
There is of course an immediate mathematical solution :   from
Theorem 2 in §1 every spacetime is paracompact and therefore admits
a Riemannian structure.   The deficiency in this solution is the
non-uniqueness of the Riemannian structure, because of the freedom
of choice available for the partition of unity on which it is based.
Another way to obtain a Riemannian structure h on a spacetime
(M,g) with a timelike vector field v is to invert the formula
of Avez [2], which we used in §1.   If we suppose that

$$g = h - \frac{2}{h(v,v)} h_\downarrow(v) \otimes h_\downarrow(v) \ ,$$

then

$$g(a,v) = -h(a,v)$$

and it follows that there is a unique Riemannian structure

$$h = g - \frac{2}{g(v,v)} g_\downarrow(v) \otimes g_\downarrow(v) \ .$$

However, this structure depends on the choice of the timelike vector
field v and so in a physical sense depends on the existence and
distinction of some matter field.   Of course many simple spacetimes
do have such fields but in general arguments, as far as practicable,
we avoid the introduction of extra structure.

Various formulations of criteria for completeness of spacetimes
have been tried (cf. Hawking and Ellis [43]) and each furnishes an
implicit definition of what is to be a singularity.   The strongest
competitor on the grounds of mathematical and physical elegance was
that of Schmidt [84] (cf. also [41], [81], [86] and [18]).   His
criterion for curves is called <u>bundle-completeness</u> or <u>b-completeness</u>.

It decides unambiguously which topologically inextensible curves
in a spacetime are finite.   The test is the length of their
horizontal lifts in the frame bundle via the metric $\hat{g}$ induced by
the Levi-Cività connection (the metric $\hat{g}$ appeared in the proof
of Theorem 5 in Ch. IV §2.8).   The great beauty of this concept
is that it uses only the connection, which physically is more
fundamental even than the Lorentz structure itself because in
principle it is directly observable through the behaviour of free
particles.   Further, a b-incomplete spacetime (M,g) can be
provided with a unique b-boundary ∂M , making MU∂M a topological
space.   The same procedure can be applied to a Riemannian manifold;
then the manifold with b-boundary coincides with the usual Cauchy
completion, giving more evidence of inherent naturality in the
procedure.   Further geometric properties of the b-completion are
discussed in [26] .

   The topology on MU∂M makes it possible to formulate questions
concerning the geometry of spacetime 'near a singularity', i.e. in
every neighbourhood of a point in ∂M .   We cannot of course expect
a spacetime structure for MU∂M because we suppose that (M,g)
itself is inextensible;   nor can we expect MU∂M to support a
manifold structure at all in general.   So it remains to enquire
what topological properties of M extend to the completion space.
It turns out that MU∂M is connected, locally connected, arcwise
connected and second countable but, unlike M , it need not be
locally compact nor more than a T_0 space.   The loss of these
properties means that a point in the b-boundary may fail to have
any compact neighbourhood and may also be inseparable (by disjoint
open sets) from other points in MU∂M.   In particular then, MU∂M
need not be a Hausdorff space and this is a serious disadvantage
for physical interpretations of singularities.

   Two different ways out of this difficulty have already been
investigated.   One uses the projective limit of a family of
b-completions of co-compact sets in M , the other uses a
parallelization to modify the metric in the frame bundle.   In
both cases the resulting completion is Hausdorff.   The price of
this improvement is on the one hand the intricacy of any actual
calculations using the projective limit or on the other hand a

restriction to parallelizable spacetimes.   In fact the latter is
not a real drawback for most physically interesting spacetimes are
parallelizable, the problem lies in choosing among many different
parallelizations.

A study of bundle completion, including detailed proofs, has
been given in a recent monograph (Dodson [26]) and we shall not
repeat that material here.   Our treatment is intended to depict
something of the universal character of both the underlying
construction and its two modifications, through the formulation
and proof of three theorems.   The first theorem shows that bundle
completion does provide a boundary for spacetime;   then the
corollaries give a breakdown of its properties, progressing from
good to bad.   Theorem 2 provides the projective limit construction
and its consequences, then Theorem 3 deals with the case of
parallelizable spacetimes.

This final section of the book is particularly open-ended
(cf. Clarke and Schmidt [18] also Tipler, Clarke and Ellis [99])
and no doubt interesting topological and geometrical results will
be found along all three lines of attack on spacetime singularities
that we describe below.   Personally, I should like next to have
some classification of parallelization completions and to know when
they coincide with other completions.   Also, in another direction,
I expect some physically interesting completions to arise from
2-jet bundles (cf. Yano and Ishihara [107] Ch. X, also Radivoiovici
[80], Hermann [45] and Hennig [44]) because of the prominent place
that second order differential equations have in physics.

5.1   Theorem

Given a spacetime (M,g) there is a unique topological space $\bar{M}$,
its b-completion, in which M is dense.   The set of completion
points

$$\partial M = \bar{M} \backslash M$$

its b-boundary, consists of endpoints for inextensible curves in
M that have finite horizontal lifts in the frame bundle metric $\hat{g}$
induced by the Levi-Cività connection.

Proof

In the presence of the pseudo-
Riemannian structure g we can
work in a connected component
O^+M of the orthonormal frame
bundle, with structure group
$SO^+(1,3)$ by Theorem 1 Corollary 1

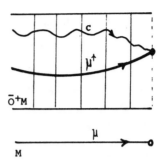

in §1.   Hence, by Theorem 5 in
Ch. IV §2.8, we use the closedness
of the Lorentz group in $G\ell(4;\mathbb{R})$
to obtain a Riemannian structure on LM and we shall denote its
restriction to O^+M by $\hat{g}$.   Also, we shall abbreviate $SO^+(1,3)$
to $\hat{O}^+$.   The construction then proceeds as follows (cf. [26] for
details).

(i)   The action of $\hat{O}^+$ on O^+M is uniformly continuous with
respect to the topological metric $d_{\hat{g}}$ determined by the
Riemannian structure $\hat{g}$.   It follows that this action has a unique
extension to the Cauchy completion metric space $(\bar{O}^+M, \bar{d}_{\hat{g}})$, and
we define $\bar{M}$ to be the quotient $\bar{O}^+M/\hat{O}^+$.

(ii) Points in $\bar{O}^+M \backslash O^+M$ are equivalence classes of Cauchy
sequences in O^+M ,   each of which is representable by a piecewise
C^1 curve

$$c : [0,1) \longrightarrow O^+M$$

with bounded length

$$\int_0^1 \hat{g}(\dot{c},\dot{c})^{\frac{1}{2}} \; < \; \infty$$

but without a continuous extension in O^+M to domain $[0,1]$.

(iii) For a curve such as c in (ii), its projection onto M ,
given by

$$\mu : [0,1) \longrightarrow M : t \longmapsto \Pi_{O^+} \circ c(t) ,$$

is likewise inextensible in M to domain $[0,1]$.   Further, given
any $u \in \Pi_{O^+} \circ \mu(0)$,   the (unique) horizontal lift $\mu^\uparrow_u$ of μ
through u has bounded length in $(O^+M,\hat{g})$.   Then $\mu^\uparrow_u$ determines
a point in the boundary $\bar{O}^+M \backslash O^+M$ and it turns out that this point

lies in the same orbit of $\hat{o}^+$ as did the endpoint of c.

(iv) The boundedness of $\mu_u^{\uparrow}$ does not depend on the choice of $u \in \Pi_{O^+}^{\leftarrow} \circ \mu(0)$ through which the lift is made.

(v) Given any $z \in \partial M = \bar{M} \backslash M$, every neighbourhood of z in the quotient topology of $\bar{M}$ contains points in M on a curve like μ in (iii). Hence M is dense in $\bar{M}$ and so ∂M is truly a boundary of M .   □

Corollary 1

A spacetime (M,g) has $\partial M = \emptyset$ if and only if $(o^+M, d_{\hat{g}})$ is Cauchy complete.   □

We say that (M,g) is <u>b-complete</u> if $\partial M = \emptyset$ and <u>b-incomplete</u> if $\partial M \neq \emptyset$. We have earlier remarked that there is a geodesically complete spacetime having incomplete other curves. So geodesic completeness is insufficient to guarantee b-completeness. However, it does work the other way.

Corollary 2

If (M,g) is b-complete then it is also geodesically complete.

Proof

If $\partial M = \emptyset$ then also $(o^+M, \hat{g})$ is complete and so all of its finite curves have endpoints in o^+M . In particular, any horizontal curve of finite length in o^+M has endpoints in o^+M . Now, it is known that every geodesic in (M,g) is the projection of an integral curve of some standard horizontal vector field. Since these integral curves must be complete, their underlying geodesics must be capable of extension to infinite parameter values.   □

Corollary 3

Every point x in spacetime has a neighbourhood E_x which is b-complete.

Proof

Choose E_x to be a subset onto which the exponential map at x is a diffeomorphism (cf. [26] p. 356 for details). Schmidt [85] has shown that the b-boundary of E_x coincides with its topological boundary.   □

Corollary 4

Plainly the process of b-completion is applicable also to Riemannian manifolds, then the b-completion coincides with the Cauchy completion.   (Equally, the process is applicable to any manifold with connection.)

Proof

If (M,g) is a connected Riemannian n-manifold then we again have an orthonormal bundle O^+M , but with structure group SO(n) , and on it a Riemannian structure $\hat{g} = \theta \cdot \theta + \omega \cdot \omega$.

For any $x \in M$, $u \in \Pi_{O+}^+(x)$ and any

$$X = X_H \oplus X_O \in T_u O^+M ,$$

we find, by definition of $\hat{g}$ (cf. Theorem 5 in Ch. IV §2.8)

$$\hat{g}(X,X) \quad = \quad \|\theta(X_H)\|^2 + \|\omega(X_O)\|^2 .$$

And since u is orthonormal in g ,

$$g(D_u \Pi_{O+} X, \ D_u \Pi_{O+} X) \quad = \quad \|\theta(X_H)\|^2 .$$

So, lengths of curves in (M,g) coincide with their horizontal lifts in $(O^+M, \hat{g})$.   Also, any other curves in the bundle O^+M between two given fibres are longer than horizontal curves.   Hence the distance between fibres in O^+M coincides with the distance between their underlying points in M.   The result follows.   □

Corollary 5

$\bar{M}$ need not be locally compact.

Proof

(Cf. Schmidt [84] and Hawking and Ellis [43] p. 283.)

We apply Corollary 4 to the Riemannian submanifold

$$M = I\!R^2 \backslash \{ (x, \sin x^{-1}) \mid x \neq 0 \} \cup \{ (0,y) \mid |y| \leqslant 1 \}$$

of two dimensional Euclidean space.   In fact, the b-boundary of $\bar{M}$, the connected component of $(0,-2) \in M$, is

$$\partial M^- = \{(x, \sin x^{-1}) \mid x \neq 0\} \cup \{(0,-1)\} \ .$$

Hence, $M^- \cup \partial M^-$ is not locally compact because $(0,-1)$ has no compact neighbourhood.   □

Corollary 6

An attempt to recover omitted points by supplying a b-boundary may yield a different space.

Proof

Consider the Riemannian submanifold of R^2 ,

$$M = R^2 \setminus \{(0,y) \mid |y| \leqslant 1\} \ .$$

For each $(0,y)$ omitted from R^2 to form M , there are two points in the b-boundary ∂M , except where $y = \pm 1$ where there is only one point.   □

The other topological problem with $\bar{M}$ is that it need not be Hausdorff. A necessary and sufficient condition for Hausdorffness is closedness of the graph of $\hat{O}^+$ in $\bar{O}^+M \times \bar{O}^+M$, where $\bar{O}^+M$ is the Cauchy completion of O^+M . The proof is given in [26] p. 433, together with details of the failure even of the T_1 separation property in $\bar{M}$ if $\hat{O}^+$ does not have closed orbits in $\bar{O}^+M$. However, more to the point for cosmology is the failure of the bundle completion of popular spacetimes to be Hausdorff.

Corollary 7

The b-completion $\bar{M}$ of a spacetime (M,g) need not be Hausdorff.

Proof

Two celebrated examples are Friedmann spacetime (with $M = R \times S^3$ representing a universe expanding from an initial Big Bang singularity) and Schwarzschild-Kruskal spacetime (representing the black hole resulting from stellar collapse). The first was discussed at considerable length in [26] and the second was studied in particular by Johnson [55] (cf. also Clarke [16]).   □

A simpler example of a non-Hausdorff b-completion is the case of $M = S^1$ with a constant connection (cf. Dodson and Sulley [29]). There the b-boundary is a single point outside S^1 and the only

open neighbourhood of this point in
the b-completion $\bar{S}^1$ is the whole
of S^1 . This example precisely
illustrates the following sufficient
condition for a bundle completion to
be non-Hausdorff. For, every curve
that circles S^1 indefinitely is
b-incomplete, has every point of S^1
as a limit point and ends on the
b-boundary. (Cf. also [26] p. 434
§3.5.)

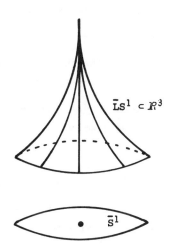

$\overline{LS}^1 \subset \mathbb{R}^3$

$\bar{S}^1$

Corollary 8

A point x in a spacetime (M,g) is not Hausdorff separated in
$\bar{M}$ from a point $y \in \partial M$ if there is an inextensible curve c in M
which has x as a limit point and y as an endpoint.

Proof

(Hawking and Ellis [43] p. 289.)

Given such a curve c then it has a horizontal lift $c^\dagger$ ending
at some b in the orbit of $\hat{O}^+$ in $\bar{O}^+M$ over y . If V is an
open set in $\bar{M}$ containing y then the bundle of orbits of $\hat{O}^+$
in $\bar{O}^+M$ over V is also open, it contains b and therefore $c^\dagger$
eventually remains in it. Hence V meets every neighbourhood of
x because x is a limit point of c . $\quad\square$

We turn now to a modification of the process of b-completion,
first suggested by C.J.S. Clarke. It was discussed at the GR8
Conference in 1977 (cf. Clarke [15]) and in a series of subsequent
papers (cf. [16], [26], [27], [91] and [99]).

5.2 Theorem

Given a spacetime (M,g) there is a unique topological space $M^\bullet$,
that is a projective limit formed from b-completions of co-compact
subsets of M , in which there is a dense subset M° that is
homeomorphic to M . The projective limit boundary

$$\partial^\bullet M = M^\bullet \backslash M^\circ$$

so determined is Hausdorff separated from M° .

Proof

We indicate the steps only, emphasising the categorical aspects of the construction (cf. [27], [91]).

(i)   Denote by T the topology on M and also put

$$T^0 = \{V \in T \mid M\backslash V \text{ is compact}\}$$

$$= \{V_\alpha \mid \alpha \in A\} ,$$

for some indexing set A . Then $T^0 \cup \{\emptyset\}$ is a topology.

(ii)   Denote by $\bar{V}_\alpha$ the b-completion of V_α and define

$$M_\alpha = (M \amalg \bar{V}_\alpha)/\sim\alpha , \quad \text{for each } \alpha \in A ,$$

to be the pushout (cf. Ch. II §2.3) of the inclusion diagram

$$M \hookleftarrow V_\alpha \hookrightarrow \bar{V}_\alpha .$$

By completeness of Top we know that M_α inherits a topology (cf. Ch. II §2.6) unique by being the largest among those that support continuity of projection from $M \amalg \bar{V}_\alpha$. Denote this topology by T_α .

(iii) The indexing set A has a partial order $\leqslant$ by

$$\alpha \leqslant \beta \iff V_\alpha \subseteq V_\beta .$$

So, for any $\alpha \leqslant \beta$ we can use the universal property of the pushout squares to obtain a unique continuous map $p_{\alpha\beta}$ by

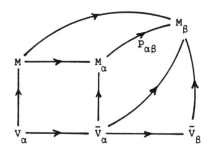

(iv)   The diagram so formed in Top :

$$\Delta = \{p_{\alpha\beta} : (M_\alpha, T_\alpha) \longrightarrow (M_\beta, T_\beta) \,|\, \alpha, \beta \epsilon A \; , \; \alpha \leqslant \beta\}$$

satisfies the requirements in Ch. III §2.8 and so we can form its projective limit space $(M^{\bullet}, T^{\bullet})$, together with a family of continuous projections

$$\chi_\alpha : M^{\bullet} \longrightarrow M_\alpha : (x_\lambda)_{\lambda \epsilon A} \longrightarrow x_\alpha$$

commuting with all

$$M_\lambda \xrightarrow[p_{\lambda\alpha}]{} M_\alpha \xrightarrow[p_{\alpha\beta}]{} M_\beta \; .$$

Note that $M^{\bullet}$ consists of certain sequences drawn from the M_α , $\alpha \epsilon A$.

(v)   Next we use the universal property of the left limit of Δ in (iv).   We know (cf. II §2.1) that given any

$$\Delta' = \{f_\alpha : K \longrightarrow M_\alpha \,|\, \alpha \epsilon A\}$$

in Top, commuting with Δ then there exists a unique continuous map

$$f : K \longrightarrow M^{\bullet}$$

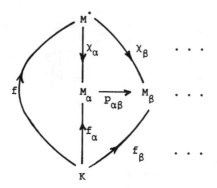

satisfying

$$f_\alpha = \chi_\alpha \circ f .$$

The first candidate for Δ' is

$$\{M \xrightarrow[f_\alpha]{} M_\alpha | \alpha \epsilon A\}$$

where, by our construction, each f_α is the homeomorphism of M onto $f_\alpha M \subseteq M_\alpha$ generated by the injection of M into $M \coprod V_\alpha$. Hence we obtain the unique continuous map

$$f : M \longrightarrow M^\bullet : x \longmapsto (x_\alpha = x)_{\alpha \epsilon A}$$

whose image consists of the constant sequences drawn from the M_α, $\alpha \epsilon A$. This f is evidently a bijection and in fact also a homeomorphism onto its image

$$M^\circ = fM ,$$

Also, M° turns out to be dense in $M^\bullet$. Therefore,

$$\partial^\bullet M = M^\bullet \backslash M^\circ$$

is indeed a fair candidate for a boundary of (M,g). Quite reasonable separation properties follow (cf. [26] for details):

If $x,y \epsilon M^\bullet$ and

<u>either</u>   (a) $\chi_\alpha(x)$ and $\chi_\alpha(y)$ are Hausdorff separable
in (M_α, T_α) for some $\alpha \epsilon A$,

or (b) $y \in M^{\circ}$,

then x and y are Hausdorff separable in $(M^{\bullet}, T^{\bullet})$. $\square$

So, $(M^{\bullet}, T^{\bullet})$ has the declared properties and we see that
intuitively separate points in $\bar{V}_{\alpha}$, the b-completion of some
co-compact $V_{\alpha} \in T$, are kept separate in $M^{\bullet}$. In particular,
interior points are separated from $\partial^{\bullet} M$. Now, a continuous curve
in M that remains in a compact set $M \backslash V_{\alpha}$ may determine a point in
the b-boundary ∂M . However, the image in $fM = M^{\circ} \backslash M^{\bullet}$ of such
a curve cannot continuously end on the projective limit boundary
$\partial^{\bullet} M$. For it too is trapped in a compact set, $f(M \backslash V_{\alpha}) \subset M^{\circ}$,
and M° is Hausdorff separated from $\partial^{\bullet} M$.

Corollary 1

(i)   Let $c : [0,1) \longrightarrow M$ be a b-incomplete curve not trapped
in (or indefinitely often returning to) a compact set, then the
curve

$$f \circ c : [0,1) \longrightarrow M^{\circ}$$

has an endpoint in $\partial^{\bullet} M$

(ii)   If $x \in \partial^{\bullet} M$ then it is the endpoint of $f \circ c$ for some
b-incomplete curve c in M that is not trapped in (or indefinitely
often returning to) a compact set.

Proof

The details are rather intricate and can be found in Slupinski
and Clarke [91]. We shall just note a further use of the universal
property of the projective limit construction (cf. (v) in the proof
of the Theorem). In (i) the given curve c eventually remains
outside every compact set $M \backslash V_{\alpha}$, so eventually it remains in V_{α}
and hence ends on ∂V_{α} . Thus we have continuous

$$c_{\alpha}^{\bullet} : [0,1] \longrightarrow M_{\alpha} , \quad (\forall \alpha \in A) ,$$

commuting with the $p_{\alpha \beta}$ and therefore also a continuous curve

$$c^{\bullet} : [0,1] \longrightarrow M^{\bullet} .$$

But $c^{\bullet}$ coincides with $f \circ c$ on $[0,1)$. $\square$

It is clear that $M^{\bullet}$ is physically more acceptable than $\bar{M}$

as a completion space for spacetimes because the usual topological
requirements for distinguishing points persist right up to the
boundary in M˙ .   There is another particular advantage, arising
from the closed Friedmann spacetimes.   These begin with an initial
Big Bang singularity, expand, then subsequently collapse into a
final singularity after a finite time.   In the b-completion the
initial and final singularities are topologically identified (full
details can be found in [26]).   This unphysical behaviour is
resolved by the projective limit completion.

Corollary 2

The initial and final singularities in closed Friedmann spacetime
are Hausdorff separable in the projective limit completion M˙ .

Proof

Topologically we have here

$$M \equiv (0,1) \times S^3$$

and geometrically,

$$g = \rho^2 \cdot (-g_\tau) \otimes g_\sigma \qquad (\text{cf. } \S4)$$

where the conformal factor is typically given by

$$\rho : (0,1) \longrightarrow (0,1) : t \longmapsto 1 - \cos 2\pi t .$$

Physically we would expect the initial and final singularities to
be separate, but in the b-completion they actually coincide.
This does not happen in the projective limit completion.   For,
removal of a compact set like

$$K_\alpha = [\tfrac{1}{4}, \tfrac{1}{2}] \times S^3$$

yields a disconnected co-compact set

$$V_\alpha = \left((0, \tfrac{1}{4}) \times S^3 \right) \cup \left((\tfrac{1}{2}, 1) \times S^3 \right) .$$

Then the initial and final singularities are Hausdorff separable
in (M_α, T_α) and therefore also in $(M˙, T˙)$, by the theorem.   ☐

(It is known (cf. [99] §5.2) that the initial and final singularities
are not separated in the projective limit completion of the Friedmann

spacetime with negative curvature and negative cosmological constant. There, a suitable disconnecting set is not available.)

Finally, we show that similar advantages over the b-completion are available if we are prepared to commit ourselves to a choice of parallelization. We noted earlier (cf. §3) that any physically reasonable spacetime is likely to be parallelizable; the crucial point is an absence of information on how to choose among different available parallelizations. Any information that effectively does distinguish one particular parallelization necessarily imparts extra structure to a spacetime. Physically, extra structure that would fix a choice of parallelization means assuming something like a matter field, which is perfectly reasonable in many situations.

The p-completion available through a parallelization p was discussed at the GR8 Conference in 1977 (cf. [25]) further developed in [26] and shown generally to yield a Hausdorff space by Dodson and Sulley [30]. The motivation for it came from the closed Friedmann b-completion, where initial and final singularities are identified through a sequence of curves that are the result of unboundedly 'high' lifts in the frame bundle. The bundle metric in the p-completion has a vertical term which makes such lifts prohibitively expensive in length and consequently the idenfication is severed. The frame bundle geometry that ensues from using this metric retains many of the good features present in the geometry of Schmidt's metric: The group action remains uniformly continuous; different inner products invoked for the canonical 1-form and connection forms yield a uniformly equivalent metric structure; fibres of the frame bundle are complete, homogeneous spaces. The geometry is studied in [26] , including an analysis of the closed Friedmann spacetime. Here we shall exploit the coincidence, on a parallelization section, of the two alternative Riemannian structures for the frame bundle; further details can be found in [30]. We also show that in at least one case of physical importance the p-completion process gives the usual topological completion when the b-completion process does not.

5.3   Theorem

Given a spacetime (M,g) and a parallelization p : M ⟶ LM
then there is a unique topological space $\widetilde{M}^p$, its p-completion,
in which M is dense.   This $\widetilde{M}^p$ is homeomorphic to the Cauchy
completion of pM as a Riemannian submanifold of $(LM,\widetilde{g}_p)$;
in consequence it is a Hausdorff space.

Proof

We know that pM is a diffeomorphic image of M in LM , since
we suppose that p is at least C^1 .   In Theorem 8 of Ch. IV
§2.8 we noted the Riemannian structure $\hat{p}$ induced on M by

$$\hat{p} = \hat{g}\circ(D_xp,\ D_xp)$$

which is, by the Corollary there, also equivalent to

$$\widetilde{g}_p\circ(D_xp,\ D_xp)\ .$$

Hence we can define $\widetilde{M}^p$ to be the Cauchy completion of the
Riemannian manifold $(M,\hat{p})$ viewed as a metric space.   Then it
necessarily contains M as a dense subset and so we do have

$$\widetilde{\partial}M = \widetilde{M}^p\backslash M$$

as a well-defined boundary of M ,   the p-boundary.

By construction, the diffeomorphism p is also an isometry
and so $(M,\hat{p})$ and $pM \subset (LM,\widetilde{g}_p)$ are isomorphic in the category of
metric spaces, therefore they have homeomorphic Cauchy completions.
Like all metric spaces, $\widetilde{M}^p$ has a Hausdorff topology.   ☐

Corollary 1

The p-completion $\widetilde{M}^p$ is homeomorphic to the quotient space $\widetilde{LM}/G$
where $\widetilde{LM}$ is the Cauchy completion of LM in the metric given by
the Riemannian structure $\widetilde{g}_p$.

Proof

The appropriate curves are constructed in [30] .   ☐

Corollary 2

The topology of the b-boundary of (M,g) may differ from that of
its p-boundary.

Proof

Take $M = (0,1) \times S^1$ and

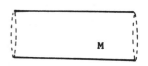

$$g = (1-\cos 2\pi t)^2 (-g_\tau) \otimes g_\sigma$$

where g_τ and g_σ are the standard
Riemannian structures on $(0,1)$ and $S^1 = \mathbb{R}(\mathrm{mod}\ 1)$,   respectively.
This is a two-dimensional submanifold of closed Friedmann spacetime,
which faithfully carries the crucial geometry - and the
'singularities' at $t = 0,1$.

There is a parallelization, orthonormal in g ,

$$p : M \longrightarrow OM : (t,\sigma) \longmapsto (\partial_t, \partial_\sigma)/(1-\cos 2\pi t) .$$

Then it turns out that :

(i)   fibres of $\bar{O}M$ over $t = 0,1$ are degenerate   (via $\hat{g}$) ;

(ii)   fibres of $\tilde{O}M$ over $t = 0,1$ are non-degenerate   (via $\tilde{g}_p$) ;

(iii) the singularities at $t = 0,1$ coincide in one point, in the
b-completion $\bar{M}$;

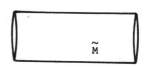

(iv)   the singularities at $t = 0,1$ are
disjoint copies of S^1 ,   in the
p-completion $\tilde{M}$ and topologically
$\tilde{M}$ is the usual completion of $(0,1) \times S^1$,
namely

$$\bar{M} \qquad$$

$$\tilde{M} = [0,1] \times S .$$

The detailed calculations can be found
in [26], based on the original
investigations by Bosshard [8] and
Johnson [55] of the b-completion of
Friedmann spacetime.   □

$$\tilde{M}$$

At first sight one might claim
that the natural completion for an open
cylinder like $(0,1) \times S^1$ is a closed
cylinder, as given by $\tilde{M}$ above.
However, that is an appeal to naturality

$(1-\cos 2\pi t)^2$

t

in the category Top - or Met, to be precise.   What might not be
adequately represented, is the collapse in the pseudo-Riemmanian
g of the geometry of space (S^1) as either singularity is
approached.   The question is really a physical one for the
Friedman spacetime : at the instant of creation from the Big Bang,
would a particle have a well-defined
sense of direction, and in the instant
of a final collapse would a particle
know from where it had come?   In the
p-completion $\tilde{M}$, spacelike directions
do persist into the singularities; but
in the b-completion $\overline{M}$ they do not and
that is viewed by Tipler, Clarke and
Ellis [99] §5.2, as unreasonable.
This defect is shared by the projective
limit completion $M^\bullet$ - though there
at least a cylinder remains a cylinder,
instead of becoming a pinched torus.
For a particle in our 2-dimensional
closed Friedmann spacetime, the
available completion spaces may seem
like choosing a destiny from among
eternal cycling, simple oblivion and
organised heaven.   If heaven is chosen,

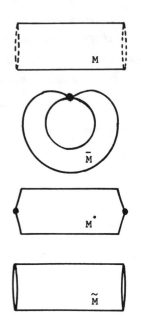

then it must be paid for in advance by the selection of a
parallelization; random reincarnation or oblivion can be had for
nothing.

6. STABILITY MATTERS

The detailed study of the spacetime singularity theorems by Hawking and Ellis [43] showed that the *occurence* of singularities is stable under perturbations of the initial Cauchy data on a compact hypersurface . However, this stability did not extend to the *character* of the singularities in general, as they demonstrated by an example (p 360) of a collapsing dust cloud. In this section we summarise some recent work on several topological and geometrical aspects of the stability and instability of completeness and incompleteness that appear to be relevant to the study of spacetime structure. Mainly, we throw more light on the criteria and stability of occurence of singularities but we offer also some new methodology that may serve in the study of their nature, and elsewhere in physical field theory.

Chapter 6 in the book of Beem and Ehrlich [109],(cf also [108]) is devoted to the study of stability properties for Robertson-Walker (i.e. Friedmann) space-times.In particular, it is shown there that the incompleteness of nonspacelike geodesics is stable under small C^2-perturbations of Robertson-Walker big bang models.A property is called stable if it holds on an open set.Here,the topology used on the space of Lorentz metrics is the Whitney W^2- or fine C^2-topology,which has, as base neighbourhoods, sets of metrics with components and their derivatives up to second order *nearby* in any chart.We give a more precise description below and indicate some results.

Williams, in his thesis [137], pointed out that the Whitney topologies may not be the most appropriate choice for spaces of connections. For example, the Levi-Civita map, which sends metrics to symmetric linear connections, is not con-tinuous with respect to Whitney topologies.On the other hand, if we give the space of connections the D^k-topology, which is algebraically sensitive, then we do have continuity of the Levi-Civita map, and of other important maps. However, on a Lorentz torus, geodesic completeness and geodesic incompleteness are unstable in both D^k and W^k-topologies.

Canarutto and Dodson [114] used the notion of a universal connection on the bundle (actually the *system*) of all linear connections, to establish the stability of b-incompleteness. This construction involves us in setting up the concept of a connection on a fibred manifold as a section of the first jet bundle and we shall describe some of the results obtained so far.A survey of work on the geometry of singularities has recently been given by Canarutto [113], providing a large number of references.

Vickers [136] studied quasi-regular singularities, for which the curvature is well-behaved in a region around the singularity. These may well not correspond to any *physical* singularity, since tidal forces will be well-behaved in their vicinity. However, they are serious for geometrically or topologically global reasons because the spacetime cannot be extended to remove them. It turns out that an important class of quasi-regular singularities are unstable.

6.1 Jet bundles

Here we outline the basic structures that we need;more details can be found in Michor [128].

A k-jet at x from a manifold X to a manifold Y is an equivalence class $j^k f(x)$ of maps $f : X \longrightarrow Y$ which have the same derivatives up to order k at x in some (hence any) chart about $x \in X, f(x) \in Y$.Note also that if $h \in j^k f(x)$ then $h \in j^l f(x)$ for $0 \leq l \leq k$.The set of all k-jets at all points of X is denoted $J^k(X, Y)$,it turns out to be a C^{r-k} fibre bundle over $X \times Y$, when X and Y are C^r for some $r \geq k$.In some particular cases it is easily identified; let M be an n-manifold:

(i) $J^1(\Re^1, M) \cong TM$, the tangent bundle to M.

(ii) $J^1(\Re^n, M)$ contains LM, the frame bundle to M as an open submanifold consisting of 1-jets at $0 \in \Re^n$ of local diffeomorphisms of open neighbourhoods of $0 \in \Re^n$.

The case (i) can be viewed as representing the tangent bundle by classes of curves *going through the same point in the same direction* . In (ii), frames of LM are faithful copies of frames from $\Re^n$ at the origin; the geometry of this is studied in detail by Cordero, Dodson and deLeon in [117].

There are obvious projections $J^{k+1}(X, Y) \longrightarrow J^k(X, Y)$ and the projective limit (cf II §2.1) for $k \geq 0$ is denoted $J^\infty(X, Y)$ and turns out to be a C^∞ fibre bundle. We call $J^k(X, Y)$ the <u>bundle of k-jets</u> over $X \times Y$. The jet approach to mechanics and field theory is described by deLeon and Rodrigues [118].

6.1.1 <u>Jets of sections</u>

In differential geometry and its applications, many of the objects of interest have representations as sections of appropriate bundles. For applications, we usually are interested in sections equivalent up to some finite order of differentiation; thus,two metric tensor fields agreeing up to first order will generate the same geodesics. The appropriate way to handle spaces of equivalence classes of differentiable maps is by means of jet bundles,which were invented by Ehresmann. Michor [128] gives a modern presentation of the theory and describes the relevant topologies.

Let $p : E \longrightarrow B$ be a (smooth) surjective submersion, then p has maximum rank everywhere; this is called a <u>surmersion</u> or a <u>fibred manifold</u>. A fibre bundle is an example of a fibred manifold but the converse is not true. For, if $p : E \longrightarrow B$ is a fibred manifold then about each point in E there is a product neighbourhood — so E is locally a product — but E need not be locally trivial. This is seen in the fibred manifold

$$p : \Re^2 \setminus 0 \longrightarrow \Re : (x, y) \mapsto x$$

because its fibres are not even homeomorphic. The interesting thing about a fibred manifold is that it is the least structure that will support the notion of a connection.

Denote by $Sec(E/B)$ the set of (smooth) sections of the fibred manifold $p : E \longrightarrow B$ and by $J^k E$ the subbundle of $J^k(B, E)$ generated by sections. In particular, $J^0 E$ is identified with E. It turns out that $J^k E$ is a (linear) affine bundle over $J^{k-1}E$, for $k \geq 1$, with associated vector bundle $\bigvee_k T^* B \otimes VTE$ where $\bigvee_k T^* B$ is the symmetrized tensor product and VTE is the vertical subbundle of TE . Further discussion of these constructions can be found in Mangiarotti and Modugno [127]. We call $J^k E$ the <u>bundle of k-jets</u> of sections of E over B .

6.1.2 <u>Theorem</u>

Let $p : E \longrightarrow B$ be a smooth vector bundle. Then:

(i) $J^k E$ is a vector bundle over B for each k.

(ii) $J^k E \hookrightarrow J^k(B, E)$ is a smooth embedding fibred over

$$(1_B, s) : B \longrightarrow B \times E \quad s \in Sec(E/B).$$

(iii) J^k is a covariant functor on vector bundles and the embedding in (ii) is a natural transformation.

Proof Michor [128] p.8. □

The case when E is a vector bundle is especially important:
The Whitney C^k-topology or W^k-topology on the space of C^r sections of a fibred manifold $p : E \longrightarrow B$ has a base consisting of all sets of the form

$$W(U) = \{ C^r f \in Sec(E/B) \mid j^k f(B) \subseteq U \}$$

where U is open in $J^k E$ .In the sequel we shall take C^r to be C^∞ ,to save making repeated qualifications. Michor [128] and Williams [137] treat the general case. The projective limit topology (cf III §2.8) for $k \geq 0$ here is actually the union topology. Observe that, for $0 \leq k \leq l$ we have $W^k \leq W^l$, in the partial order for topologies (cf III §2.1); so the fineness of the topology increases with the order of the jets used.

6.2 Connections as sections of jet bundles

Now, we shall be particularly interested in bundles of 1-jets, so we shall agree to write JE for $J^1 E$.

A connection on a fibred manifold $p : E \longrightarrow B$ is a section of the first jet bundle $J^1 E \longrightarrow E$. Locally it appears in the form

$$\Gamma : E \longrightarrow JE \quad \subset \quad T^* B \otimes_E TE$$

$$: (x^\lambda, z^i) \mapsto (x^\lambda, z^i, dx^\lambda \otimes \partial_{x^\lambda} - \partial_{x^\lambda} \gamma^j dx^\lambda \otimes \partial_{z^j})$$

where $\gamma \in Sec(E/B)$ represents the class of sections determined by Γ at $e \in E$ and e has coordinates (x^λ, z^i). Observe that Γ is projectable onto 1_{TB} which appears as $dx^\lambda \otimes \partial_{x^\lambda}$.

The <u>horizontal</u> subspace $H_\Gamma \subset TE$ determined by Γ is given by the splitting (cf IV §2.6)

$$T_\Gamma = H_\Gamma \oplus VTE$$

$$: (X^\lambda, Z^i) \mapsto (X^\lambda, X^\lambda \partial_{x^\lambda} \gamma^i) \oplus (0, Z^i - X^\lambda \partial_{x^\lambda} \gamma^i)$$

and VTE is the <u>vertical</u> subspace.

To each connection Γ there corresponds its <u>connection form</u>, the vertical vector 1-form (cf IV §2.6)

$$\omega_\Gamma : E \longrightarrow T^*B \otimes_E TE$$

$$: (x^\lambda, z^i) \mapsto (x^\lambda, z^i, dz^i \otimes \partial_{z^i} - \partial_{x^\lambda} \gamma^i dx^\lambda \otimes \partial_{z^i}).$$

6.2.1 <u>Spaces of linear connections</u>

Linear connections on a manifold M are given by sections of JTM, like

$$\Gamma : TM \longrightarrow JTM \subset T^*M \otimes_{TM} TTM$$

$$: (x^\lambda, \dot{x}^\lambda) \mapsto (x^\lambda, \dot{x}^\lambda, dx^\lambda \otimes \partial_{x^\lambda} - \Gamma^\mu_{\lambda\nu} \dot{x}^\nu dx^\lambda \otimes \partial_{\dot{x}^\mu}).$$

Here $(x^\lambda, \dot{x}^\lambda)$ are coordinates on TM induced by (x^λ) on M, and the $\Gamma^\mu_{\lambda\nu}$ are the Christoffel symbols. Thus, we see that the linearity of the connection is precisely the condition for the section γ in the general definition to have local form:

$$\gamma : (x^\lambda) \longrightarrow (x^\lambda, K^\lambda_\mu \dot{x}^\mu \partial_{x^\lambda}), \ with \ \partial_{x^\nu} K^\lambda_\mu = \Gamma^\mu_{\lambda\nu}.$$

There is an equivalent formulation of linear connections as invariant connections on the principal bundle $LM \longrightarrow M$ of linear frames. Hence they appear as sections of $JLM \to LM$ which are invariant under $G = Gl(n, \Re)$.

From Garcia [124] it follows that JLM is a principal G-bundle over JLM/G, and actually an isomorph of the fibred product of JLM/G and $LM/G = M$ (cf II §2.3). Now, (cf Canarutto and Dodson [114]) JLM/G is a linear affine bundle over

M , with associated vector bundle $T^*M \otimes VLM/G$,and VLM/G is isomorphic to $M \times \xi$ where ξ is the Lie algebra of G . In fact

$$JLM/G \cong ((1_{T^*M} \otimes Tp)^{-1}1_{TM})/G \subset T^*M \otimes_{TM} TLM/G$$

where we use 1_{TM} to denote the identity section in

$$T^*M \otimes TM \subset T^*M \otimes TTM \subset T^*M \otimes TLM.$$

Here the same linear connection Γ appears as a section

$$\Gamma : LM \longrightarrow JLM$$

$$: (x^\lambda, b^\lambda_\nu) \mapsto (x^\lambda, b^\lambda_\nu, dx^\lambda \otimes \partial_{x^\lambda} - \Gamma^\mu_{\lambda\nu} b^\nu_r dx^\lambda \otimes \partial_{b^\mu_r})$$

where the $(x^\lambda, b^\lambda_\nu)$ are coordinates on LM induced by x^λ on M.
Now the linearity corresponds to the following form for $\gamma \in Sec(LM/M)$

$$\gamma : (x^\lambda) \mapsto (x^\lambda, \{K^\lambda_\mu \partial_{x^\lambda} \mid \mu = 1, 2, \ldots, n\})$$

where (K^λ_μ) is a nonsingular matrix, that is, a member of $Gl(n, \Re)$,   and $\partial_{x^\nu} K^\lambda_\mu = \Gamma^\mu_{\lambda\nu}$ as before.

6.2.2. Topologies for spaces of connections and metrics

The set $Con(M)$ of all linear connections on M can be given any Whitney W^k–topology (cf 6.1.1) for $k \geq 0$, since $Con(M)$ consists of sections of JLM over LM or,equivalently, of sections of JTM over TM . Similarly, the set $Met(M)$ of metric tensor fields on M consists of sections of $\vee_2 T^*M$ over M and admits W^k–topologies. Williams [137] showed that the Levi-Civita map sending a metric to its unique symmetric and compatible connection (cf IV §2.8.2) is not continuous with respect to these topologies. He suggested using instead the D^k–topology on $Con(M)$; it has basic open sets of the form

$$N_\Gamma = \{\Gamma' \in Sec(JTM/TM) \mid (\Gamma - \Gamma') \in N\}$$

where N is a neighbourhood of $O \in Sec(T^1_2 M/M)$ in the W^k–topology. This exploits the fact that the difference of two connections is a $\binom{1}{2}$-tensor field.
The space $Con(M)$ actually has a linear affine structure with associated vector space $Sec(T^1_2 M/M)$ hence the latter becomes a topological module in the

W^k–topology. This structure coinduces the D^k–topology on $Con(M)$. Full details are given in [137].

Denote by $SymCon(M)$ the set of all symmetric linear connections on M .Then we have :

6.2.3. Theorem (Williams [137])

Let $Con(M)$ have the D^k-topology and let $Met(M)$ have the W^k-topology, then the following are continuous:

The Levi-Civita map: $Met(M) \longrightarrow SymCon(M)$ (cf IV §2.8.2)

The Symmetrization map: $Con(M) \longrightarrow SymCon(M)$

The Parallelization map: $Sec(LM/M) \longrightarrow Con(M)$ (cf IV §2.6.4)

The Product map: $Met(M) \times Met(N) \longrightarrow SymCon(M \times N)$.

Moreover, if M is compact then $SymCon(M)$ is simply-connected. $\square$

With the same hypotheses we have (cf [119])

Corollary

The following are continuous:

The Connection-Metric map: $Con(M) \longrightarrow Met(LM)$ (cf IV §2.8.4)

The Metric Lift map: $Met(M) \longrightarrow Met(LM) \longrightarrow Con(LM)$. $\square$

6.3 D^k- and W^k-stability

We say that a property is T-stable if it holds on an open set in the T-topology,otherwise it is T-unstable.

Since, for $0 \leq k \leq l$, the W^k–topology is less fine (ie smaller) than the W^l–topology, it follows that also:

$$W^k - stable \Rightarrow W^l - stable \quad and \quad D^k - stable \Rightarrow D^l - stable$$

$$W^l - unstable \Rightarrow W^k - unstable \quad and \quad D^l - unstable \Rightarrow D^k - unstable.$$

6.3.1. Theorem

The following are D^k–stable in $Con(M)$

(i) Having nonzero torsion.

(ii) Being incompatible to a fixed metric.

Proof
(i) Symmetrization is a coninuous retraction and so its image is closed, hence its complement is open.
(ii) Apply the Levi-Civita map to the given metric; take the inverse image by symmetrization and obtain a closed set; its complement is the set of incompatible connections. □

We recall in passing, the results on *stable causality* in Hawking and Ellis [43] which characterise the stability of having no closed timelike curves in the space of metrics (cf §2.9 and §3.3 above).

6.3.2.Lorentz torus

Williams [137] constructed the family

$$g_n(x,y) = \begin{pmatrix} 0 & 1 \\ 1 & n^{-1}\sin x \end{pmatrix} \quad n = 1, 2, \ldots$$

of Lorentz metrics on the torus $\Re^2/Z \times Z$. He showed that these have incomplete null geodesics but the (g_n) converge uniformly in the W^k–topology to a complete metric as $n \longrightarrow \infty$.Next he considered the family

$$g_n(x,y) = \begin{pmatrix} 0 & 1 \\ 1 & n^{-1} + 1 - \cos x \end{pmatrix} \quad n = 1, 2, \ldots$$

These are all complete Lorentz metrics but they converge uniformly to a null incomplete metric as $n \longrightarrow \infty$. He produced other models doing the same for timelike and for spacelike geodesics. In summary then, on the torus with Lorentz metrics, geodesic completeness and geodesic incompleteness are both W^k–unstable. The torus is a particulaar case of a product of compact manifolds; the following is useful in generating more examples.

6.3.3 Theorem (Williams [137])

Let M and N be compact manifolds and $k \geq 0$. Then we have:

(i) Geodesic completeness is D^k–unstable on $M \times N$ if it is D^k–unstable on M.

(ii) Geodesic incompleteness is D^k–unstable on $M \times N$ if it is D^k–unstable on M and on N.

(iii) Lorentz geodesic completeness is W^k–unstable on $M \times N$ if it is W^k–unstable on $M \times N$. □

The examples constructed by Williams contradict the previous view (cf Lerner [126]) that, for all Lorentzian manifolds, geodesic completeness would be W^2–stable.

Beem and Ehrlich [110] obtained criteria for the stability of geodesic completeness for pseudo-Riemannian manifolds by imposing conditions on the background metric. To this end, they defined a manifold (M, g) to have a pseudoconvex null geodesic system if, given any compact set K there is a compact set K' such that any geodesic segment with both endpoints in K must lie in K'. Also, (M, g) is said to have no imprisoned null geodesics if each inextensible null geodesic fails to be contained in any compact set.

Both of these properties have been used in the study of pseudodifferential operators (cf Beem and Parker [111], Lerner [126] and Treves [134]). Here they yield some interesting results for us:

6.3.4 Theorem (Beem and Ehrlich [110])

Let (M, g) be a pseudo-Riemannian manifold with all null geodesics complete and a pseudoconvex null geodesic system with no imprisoned null geodesics. Then (M, g) is W^1–stably complete in the space of pseudo- Riemannian metrics.

The result holds also for non-spacelike and for non-timelike geodesic completeness.
□

Corollary

Null and non-spacelike geodesic completeness are both W^1−stable for globally hyperbolic spacetimes, because these have pseudoconvex geodesic systems without imprisonment.   □

6.3.5 Theorem (Beem and Parker [112])

The joint condition of having a pseudoconvex geodesic system and no imprisonment is W^1−stable, but neither condition is stable alone.   □

For conformal perturbations we have the following:

6.3.6 Theorem (Beem and Ehrlich [110])

Let (M, g) be a pseudo-Riemannian manifold. Then null geodesic completeness and null geodesic incompleteness are both W^0−stable in the space of metrics conformal to g. If M is compact, then all metrics conformal to g are either null geodesically complete or all are null geodesically incomplete.   □

In 6.7.1 below we obtain another conformal stability result, for b-incompleteness; the method there eends on the following construction.

6.4 Systems of connections

In 6.2 we looked at the general definition for a connection on a fibred manifold. Here we describe some results in Canarutto and Dodson [114] which give another way to approach stability criteria for incompleteness, by using a *system of connections*.

We shall begin with the definition for the general case of a fibred manifold $E \longrightarrow B$, our application to spacetime geometry uses the frame bundle $\pi : LM \longrightarrow M$.

A <u>system of connections</u> on a fibred manifold $p : E \longrightarrow B$ is a pair (C, η)

where $p_C : C \longrightarrow B$ is a fibred manifold and

$$\eta : C \times_B E \longrightarrow JE$$

is a fibred morphism over E.

This definition originates with Modugno, the earlier terminology being a structure of connections (cf Mangiarotti and Modugno [127]); he gives a detailed study in [129]. We call C the space of connections for the system, because there is a bijective correspondence between sections $\bar{\Gamma}$ of $p_C : C \longrightarrow B$ and connections Γ on $p : E \longrightarrow B$ (cf 6.2). We shall say that a property of a connection Γ in a system is system-stable if there is an open set V, with

$$(\bar{\Gamma} \circ p, 1_E)(E) \subseteq V \subseteq C \times_B E,$$

such that the property holds for all connections Γ' whose corresponding system sections lie in V, that is,

$$(\bar{\Gamma}' \circ p, 1_E)(E) \subseteq V.$$

6.4.1 Theorem

The system of all linear connections on a manifold M has system space

$$p_C : JLM/G \longrightarrow LM : (x, [j\gamma]) \mapsto (x, \gamma(x))$$

with $G = Gl(n, \Re)$ and the fibred morphism

$$\eta : JLM/G \times LM \longrightarrow JLM : T^*M \otimes_{LM} TLM$$

$$: ((x, [j\gamma], b) \longrightarrow \{T\gamma : T_x M \longrightarrow T_{\gamma \circ x(b)} LM\}.$$

Proof From the results of García [124] for principal bundles, it follows that there is a bijective correspondence between $Con(M)$ and $Sec((JLM/G)/M)$. Now, given $(\bar{\Gamma} \in Sec((JLM/G)/M)$, the corresponding connection is

$$\Gamma = \eta \circ (\bar{\Gamma} \circ \pi, 1_{LM}) : LM \longrightarrow JLM.$$

Also, η is evidently fibred over LM. $\square$

The structures involved in this system are described by the following, which is straightforward to prove.

6.4.2 <u>Theorem</u>

In the system of linear connections on M, JLM/G is a linear affine bundle with associated vector bundle $T^*M \otimes_{TM} VLM/G$. Moreover, VLM/G is isomorphic to $M \times \xi$, where ξ is the Lie algebra of G, and one such isomorphism is fixed for each choice of frame in LM.   □

6.5 <u>Universal connections</u>

Now we obtain the <u>universal connection</u> Λ .

6.5.1 <u>Theorem</u>

For the system of linear connections on M , the trivial bundle

$$\pi_1 : JLM/G \times_M LM \longrightarrow JLM/G$$

is a principal G-bundle and on it there is a principal connection Λ. The connection Λ is universal in the sense that every principal connection Γ (i.e. linear connection on M) is a pullback of Λ ; moreover, the corresponding connection and curvature forms of Γ are pullbacks of those of Λ .

<u>Proof</u> Details can be found in García [124]. We need in the sequel the connection Λ and its connection form. Explicitly,

$$\Lambda : JLM/G \times_M LM \longrightarrow J(JLM/G \times_M LM) \subseteq$$

$$\subseteq \ T^*(JLM/G) \otimes_{TLM} T(JLM/G \times_M LM)$$

$$: (x^\lambda, \gamma^\lambda_{\mu\nu}, b^\mu_\nu) \mapsto \{ \ (X^\lambda, Y^\lambda_{\mu\nu}) \mapsto (X^\lambda, Y^\lambda_{\mu\nu}, b^\mu_\nu \gamma^\lambda_{\mu\theta} X^\theta) \ \}$$

and its connection form is the vertical vector 1-form on the base JLM/G given by

$$\omega_\Lambda : JLM/G \longrightarrow T^*(JLM/G \times_M LM) \otimes_{TLM} V(JLM/G \times_M LM)$$

$$: (x^\lambda, \gamma^\lambda_{\mu\nu}) \mapsto \{ \ (X^\lambda, Y^\lambda_{\mu\nu}, B^\lambda_\mu) \mapsto (0, 0, B^\lambda_\mu - b^\mu_\nu \gamma^\lambda_{\mu\theta} X^\theta) \ \}$$

The universality can be seen by taking any linear connection Γ , corresponding to a section $\bar\Gamma \in Sec((JLM/G)/M)$. We observe that it injects into $JLM/G \times_M LM$

a copy S_Γ of LM as the graph of $\hat{\Gamma}$ (cf [114], [119]).
On this section S_Γ, Γ agrees with Λ ; namely,

$$\Gamma = \Lambda_{|(\Gamma \circ \pi, 1_{LM})LM} = \Lambda_{|S_\Gamma}$$

and the coordinates of the section are the Christoffel symbols of Γ.   □

Universal linear connections are discussed further in Cordero, Dodson and deLeon [117], Dodson [120] and Modugno [129], [130]. The original idea seems to have come from Narasimhan and Ramanan [131] , [132,] who like García [124] used them to prove Weil's theorem:
the characteristic cohomology classes of a principal G-bundle, determined by the substitution of the curvature form of a connection into the invariant polynomials of G, are independent of the connection used.

6.6 Geometry of the system of linear connections

It seems likely that the theory of universal connections on principal bundles should have applications in gauge theory and elsewhere, whenever a geometrical framework is needed to study families of connections simultaneously. The beauty of the framework lies in the fact that we can use connection geometry to study geometries of connections. Moreover, it extends to the much more general context of *systems of fibred manifold connections,*so opening up potential use in theories of partial differential equations.

We turn now to see it at work on the geometry of spaces of linear connections, based on results from Canarutto and Dodson [114].

6.6.1 Theorem

Every linear connection Γ on M determines a unique Riemannian metric, g_Γ on LM, and the section $\hat{\Gamma} \in Sec((JLM/G)/M)$ corresponding to Γ embeds (LM, g_Γ) isometrically as a Riemannian submanifold of $(JLM/G \times_M LM$.

Proof The metric $g_\Gamma = \hat{g}$ was described in IV §2.8.5 and 5.1 above.
On $(JLM/G \times_M LM)$ we construct an analogous symmetric bilinear form from

the universal connection Λ ,

$$f_\Lambda = \omega_\Lambda \bullet \omega_\Lambda + \Theta \circ T\pi_2 \bullet \Theta \circ T\pi_2$$

where ω_Λ is the connection form of Λ , Θ is the canonical 1-form of LM , π_2 is the canonical projection of $JLM/G \times_M LM$ onto LM, and $\bullet$ is the standard Euclidean inner product.

Evidently , f_Λ is degenerate on fibres of $T\pi_2$. However, this is irrelevant to the slice through the fibres by the embedding

$$(\bar{\Gamma} \circ \pi, 1_{LM})LM = S_\Gamma \subseteq JLM/G \times_M LM,$$

on which f_Λ restricts to be a positive definite metric , f_Γ .

Moreover, (S_Γ, f_Γ) is isometrically isomorphic to $(LM, \hat{g})$. $\square$

Some sketches of these spaces and slices like S_Γ are given in [114] for $M = \Re$ (cf also DelRiego and Dodson [119]).

6.7 System stability of b-incompleteness

We collect a number of results concerning the stability of connection - singularities represented by b-incompleteness. Singularities of this kind were introduced in 6.5 above, and include the most likely examples of physical singularities in general relativistic spacetime. A detailed study of the relevant geometry was given in Dodson [26] and updated recently by Canarutto [113].

Recall that a manifold M with a linear connection Γ is called b-incomplete if (LM, g_Γ) is incomplete as a Riemannian manifold (cf §6.6.1 and §5), though a better terminology might be Γ-incomplete. Our main result shows that this property is *stable* in the system of linear connections.

6.7.1 Theorem

For any manifold, b-incompleteness is system-stable.

Proof We suppose that M is b-incomplete with respect to a linear connection Γ. It is necessary to find (cf 6.4) an open set V with

$$(\bar{\Gamma} \circ \pi, 1_{LM})LM \subset V \subseteq (JLM/G \times_M LM)$$

and such that, for every $\Gamma' \in Sec((JLM/G)/M)$ with

$$(\Gamma' \circ \pi, 1_{LM})LM \ \subseteq \ V,$$

M is b-incomplete with respect to the connection Γ'.

In [114] we give two ways to construct suitable V ; the essentials of the methods are as follows:

(i) Take any connection Γ^0 such that Γ is Γ-horizontal and construct a continuous family

$$\{\Gamma^\epsilon \in Sec((JLM/G)/M) \ | \ \epsilon \in \Re\}$$

uniformly convergent to Γ as $\epsilon \to 0$, with respect to the metric g_{Γ^0}.

Then, for some $r > 0$, M is b-incomplete with respect to Γ^ϵ for all $| \ \epsilon \ | < r$.

(ii) Suppose that Γ is the Levi-Civita connection of a metric g on M . Let

$$\{g_\epsilon = \phi_\epsilon g_0 \ | \ \epsilon \in \Re\}$$

be a family of metrics conformal to $g = g_0$, with conformal factor ϕ_ϵ tending to 1 as $\epsilon \to 0$.

If $c : [0,1) \longrightarrow M$ is b-incomplete with respect to Γ and $\phi \circ c : [0,1) \longrightarrow \Re$ is uniformly convergent to 1 as $\epsilon \to 0$,

then, for all sufficiently small $| \ \epsilon \ |$, c is b-incomplete with respect to Γ^ϵ induced by g_ϵ .

This follows from the facts that

$$g_{\Gamma^\epsilon} = \omega_{\Gamma^\epsilon} \bullet \omega_{\Gamma^\epsilon} + \Theta \bullet \Theta$$

$$\omega_\Gamma : T_{b_\rho^\mu} LM \longrightarrow \xi l(n, \Re) \ \cong \ \Re^{n^2}$$

$$: (x^\lambda, b_\rho^\mu, \dot{x}^\lambda, \dot{b}_\rho^\mu) \mapsto (\dot{b}_\sigma^\lambda + \Gamma_{\mu\nu}^\lambda b_\sigma^\mu \dot{x}^\nu)(b_\lambda^\rho)^{-1},$$

$$(\Gamma^\epsilon)_{\mu\nu}^\lambda = \Gamma_{\mu\nu}^\lambda + \delta_\mu^\lambda \partial_\nu \log \phi_\epsilon + \delta_\nu^\lambda \partial_\mu \log \phi_\epsilon - g_{\lambda\nu} g^{\lambda\rho} \partial_\rho \log \phi_\epsilon.$$

The three terms depending on ϵ are bounded on $c[0,1)$ for sufficiently small $| \ \epsilon \ |$, so c has finite b-length in g_{Γ^ϵ}. $\square$

This theorem allows us to view connections by means of the Riemannian *slices* they generate in the system. Nearby connections generate nearby slices and any b-incompleteness persists into the immediate vicinity of a slice. Recall (cf §5) that geodesic incompleteness implies b-incompleteness, and that this is important for the study of singularities in spacetime. Quite general theorems (cf Hawking and Ellis [43]) suggest that, not only general relativity but *any* theory in which gravity is *attractive,* will predict singularities, and that these are likely to appear as b-incompleteness. Such classical singularities might be expected to be removed by an appropriate quantum theory of gravity. For the particular theory of geometric quantization (cf Simms and Woodhouse [133]), Gotay and Isenberg [125] showed that a massless Klein-Gordon scalar field on a positively curved spacetime cannot escape the collapse of its state vector. Our theorem 6.7.1 says that if one connection causes a physical singularity then so too will its small perturbations, suggesting that *no* quantum theory of gravity could remove the singularity.

It might be interesting to try to formulate the conservation of mass as a stability of singularities at its points of existence. The geometry of the embedding of the slices, in the vicinity of a slice with a singularity, may offer a way to study the characterisation of singularities and the instability of their nature. For example, how quasi-regular singularitiees are perturbed into curvature singularities (cf 6.7.3 below) and how collapsing charged dust can be perturbed between two singular states (cf Hawking and Ellis [43] p 360).

Canarutto and Michor [115] used the notion of essential b-incompleteness, namely completeness arising from curves not trapped in any compact set, to obtain the following correspondence with Whitney stability. Consider $Sec((JLM/G)/M)$ with the W^0-topology.

6.7.2 Theorem

Essential b-incompleteness is W^0-stable.

Proof Suppose that M is b-incomplete with respect to the connection Γ . Let $c : [0,1) \longrightarrow LM$ be a curve giving rise to a point z in the boundary of LM , such that the projection of z is a b-boundary point of essential b-incompleteness in M.

Construct the ball B of radius ρ , centre z , in $L\bar{M}$, the completion of (LM, g_Γ). We can arrange to have $c(t) \in B$ for all sufficiently small $(1 - t)$. Now, it is possible to construct an open submanifold

$$V \subset (JLM/G \times_M LM)$$

such that :

(i) $(\bar{\Gamma} \circ \pi, 1_{LM}) \circ c([0,1)) \subset V$

(ii) $V \cap S_\Gamma$ is $f_{\Gamma'}$-bounded in $(S_{\Gamma'}, f_{\Gamma'})$ for all Γ' such that (cf 6.5.1)

$$\pi_2(V \cap S_\Gamma) = \pi_2(V)$$

Then the following maps are W^0-continuous:

$$Sec((JLM/G)/M) \longrightarrow C^0([0,1), JLM/G) : \bar{\Gamma}' \mapsto \bar{\Gamma}' \circ c$$

$$\Phi : Sec((JLM/G)/M) \longrightarrow V \cong C^0([0,1), V) : \bar{\Gamma}' \mapsto (c, \bar{\Gamma}' \circ c)$$

Hence $\Phi^{-1}(V)$ is W^0-open.

For all $\bar{\Gamma}' \in \Phi^{-1}(V)$, the Γ'-horizontal lift of curve c has finite length in $(LM, g_{\Gamma'})$,is not trapped in any compact set and therefore c determines a point in the essential b-completion of M with respect to Γ. □

It remains to be seen whether a Whitney topology, or one arising from connection-induced embeddings in a system, is more useful in physical interpretations of stability for our *geometrical* singularities. Throughout,we have given prominence to the role of a connection as the fundamental geometrical entity. Now,from the physical point of view, it is a field controlled by equations involving other fields also; and appropriate criteria for stability of *physical* singularities may well need to be sensitive to this.

6.7.3 Instability of quasi-regular singularities

Physically, a timelike b-incomplete curve in a spacetime (remember we always presume our spacetimes are *inextensible,* p.160) indicates the presence of a curve which ends after a finite period. If, in a parallely propagated frame the curvature is unbounded along such a curve, then we have a curvature singularity with evidently serious physical implications. On the other hand, the curvature may be well-behaved along the b-incomplete curve; then we have a quasi-regular singularity. The classification of singularities of both types has been studied by Clarke and Schmidt [18] and by Ellis and Schmidt [122], [123] (cf also Clarke and Krolak [116]. In his thesis, Vickers [135] made a detailed study of quasi-regular singularities and reported [136] on their instability with respect to perturbations of the Cauchy data.

Intuitively, the geometry of a simple quasi-regular singularity is conveyed by an ordinary cone in $\Re^3$; it is an incomplete smooth manifold inextensible to the apex, but has everywhere zero curvature.
A *conical* ,locally flat spacetime can be made from Minkowski space. In polar coordinates, remove the points on $r = 0$; then identifications in the covering space generate a 2-plane of quasi-regular singularities on $r = 0$ (cf Ellis and Schmidt [122]). However, these quasi-regular singularities are not stable under perturbations. For, arbitrarily close to the *conical* data, on the surface $t = 0$ Cauchy data exists which evolves into a spacetime with a *curvature* singularity. The topology used on the space of Cauchy data is a modified Sobolev space topology and we refer to Vickers [136] for details of his constructions. We draw attention also to §7.6 in Hawking and Ellis [43] which discusses the Cauchy stability theorem.

6.7.4 Concluding remarks on Systems of connections and stability

We have now several geometrical ways to study the stability of kinds of singularities that may be expected in spacetime. They take advantage of the central role played by a linear connection for general relativity. We can see how this could be adapted to study singularities of gauge fields, by using other systems of connections on appropriate principal bundles. Then it would be interesting to see how singularities behave, between coupled fields represented by or associated with connections in bundles over the spacetime.

One way to view a general connection on a fibred manifold $E \longrightarrow B$ is as a lift of differential operators from TB to TE , while preserving projectability onto the identity in TB . Now, this is a very general process indeed, admitting representations of field theories of many kinds. Then systems of connections provide a convenient framework for studies of families of fields and theories, and for the formulation of criteria for stability, by exploiting the universal connections.

Dodson and Modugno [121] (cf also [130]) devised a *universal calculus* that may give an appropriate methodology for studying gauge fields; it seem likely also to yield some interesting universal cohomological results. Indeed, the Frölicher-Nijenhuis graded Lie algebra bracket admits an extension to a *universal* graded Lie algebra bracket on vector-valued forms paramatrized by jets of connections from a system on any given fibred-manifold. The universal bracket determines a *universal differential* behaving appropriately with respect to the universal connection and its curvature form. In the presence of a metric on the base manifold, we obtain a *universal codifferential* and a Laplacian. Systems of connections *are* fibred manifolds and admit distinguished systems of connections on themselves, the *overconnections* . Sections of the overconnection system space project onto sections of the base system space. These developments were motivated by the desire to handle the geometry of spacetime connections as a free field and to provide for a study of the space of Lorentz metrics through the universal calculus. They may also offer further machinery to study questions of stability in a general context. For the universal differential and codifferential operators allow gauge equations for any system to be written canonically. Moreover, Lagrangians which depend on jets of the connection seem to have real significance and the gauge-invariant Lagrangians are precisely those whose Euler-Lagrange equations admit a representation through the universal differential and codifferential.

To offer another line of potential development, we note that our connections have been all of *first order*, namely, sections of $JE \longrightarrow E$. A $\underline{k^{th}\text{-order connection}}$ on a fibred manifold $E \longrightarrow B$ is a section of $J^kE \longrightarrow J^{k-1}E$, recalling that $J^0E = E$. However, there is a fibred embedding over B of J^kE into $J(J^{k-1}E)$ and so a kth-order connection on $E \longrightarrow B$ admits interpretation as a first order connection on $J^{k-1}E \longrightarrow B$. All of the usual connection geometric objects are available for higher order connections, and they too can be organized into systems (cf Mangiarotti and Modugno [127]) with corresponding universal connections.

Bibliography

1. Arbib, M.A. and Manes, E.G. (1975) <u>Arrows Structures and</u>
    <u>Functors</u> Academic Press, New York.

2. Avez, A. (1963) "Essais de géometrie Riemannienne
    hyperbolique globale." Ann.Inst.Fourier 132,
    105-190.

3. Bass, R.W. and Witten, L. (1957) "Remark on cosmological
    models." Rev.Mod.Phys.29, 452-453.

4. Binz, E. and Herrlich, H. (1976) Eds. <u>Categorical Topology</u>
    Springer-Verlag, Berlin.

5. Birkhoff, G. and Bartree, T.C. (1970) <u>Modern Applied Algebra</u>
    McGraw-Hill, New York.

6. Bishop, R.L. and Crittenden, R.J. (1964) <u>Geometry of Manifolds</u>
    Academic Press, New York.

7. Bishop, R.L. and Goldberg, S.I. (1968) <u>Tensor Analysis on</u>
    <u>Manifolds</u> MacMillan, New York.

8. Bosshard, B. (1976) "On the b-boundary of the closed
    Friedmann model." Common.Math.Phys. 46,
    263-8.

9. Brickell, F. and Clark, R.S. (1970) <u>Differentiable Manifolds</u>
    Van Nostrand, London.

10. Cartan, E. (1966) <u>Theory of Spinors</u> MIT Press, Cambridge.

11. Cartan, H. (1971) <u>Differential Forms</u> Kershaw, London.

12. Chevalley, C. (1946) <u>Theory of Lie Groups</u> Princeton
    University Press, Princeton.

13. Clarke, C.J.S. (1970) "On the global isometric embedding of
    pseudo-Riemannian manifolds" Proc.Roy.Soc.
    A314, 417-428.

14. Clarke, C.J.S.   (1971) "Magnetic charge, holonomy and
                              characteristic classes," GRG 2, 43-51.

15. Clarke, C.J.S.   (1977) "Boundary definitions" Contribution
                              at GR8 Conference; to appear in GRG (1979).

16. Clarke, C.J.S.   (1978) "The singular holonomy group" Commun.
                              Math.Phys. 58, 291-297.

17. Clarke, C.J.S.   (1979) Elementary General Relativity, Edward
                              Arnold, London.

18. Clarke, C.J.S. and Schmidt, B.G.   (1977) "Singularities: the
                              state of the art." GRG 8, 129-137.

19. Crumeyrolle, A.   (1969) "Structures spinorielles" Ann.Inst.H.
                              Poincaré A11, 19-55.
                              (Continued in: "Groupes de spinorialité"
                              Ann.Inst.H.Poincaré A14 (1971) 309-323.)

20. Császár, A.   (1978) General Topology Adam Hilger, Bristol.

21. Devlin, K.J.   (1977) The Axiom of Constructibility Springer-
                              Verlag, Berlin, Lecture Notes in Mathematics
                              617.
                              (Cf. also Fundamentals of Contemporary Set
                              Theory, Springer-Verlag New York 1979)

22. DeWitt, C. and Wheeler, J.A.   (1968) Eds: Battelle
                              Rencontres in Mathematics and Physics
                              Benjamin, New York.

23. DeWitt, C. and DeWitt, B.   (1964) Eds. Relativity, Groups
                              and Topology Blackie and son, London.

24. Dodson, C.T.J.   (1974) "On the Cartan structural equations
                              for the local curvature and torsion of smooth
                              manifolds." Matrix Tensor Q. 25, 1, 20-32.

25. Dodson, C.T.J.   (1977) "A new bundle completion for
                              parallelizable spacetimes" Contribution at
                              GR8 Conference; to appear in GRG (1979).

26. Dodson, C.T.J.   (1978) "Space-time edge geometry" Int.J.Theor.
                              Phys. 17, 6, 389-504.

27. Dodson, C.T.J.   (1979) "Spacetime projective boundaries."
                              Contribution at Bolyai Society Conference;
                              to appear in Proceedings (1980).

28. Dodson, C.T.J. and Poston, T. (1979) Tensor Geometry Pitman,
             London (2nd edition).

29. Dodson, C.T.J. and Sulley, L.J. (1977) "The b-boundary of S^1
             with a constant connection." Lett.Math.Phys.
             1, 301-307.

30. Dodson, C.T.J. and Sulley, L.J. (1980) "On bundle completion
             of parallelizable manifolds." Math.Proc.
             Camb.Phil.Soc. (in press).

31. Eells, J. (1974) "Fibre bundles" in Global Analysis and
             its Applications Vol 1 IAEA, Vienna,
             pp.53-82.

32. Ehresmann, C. (1951) "Les connexions infinitésimales dans
             une espace fibré differentiable." Colloque
             de Topologie, Bruxelles, 5-8 June 1950,
             CBRM (Georges Thone, Liege) pp. 29-55.

33. Flanders, H. (1963) Differential Forms Academic Press,
             New York.

34. Friedrich, H. (1974) "Construction and properties of space-
             time b-boundaries." GRG 5, 681-697.

35. Gel'fand, I.M., Minlos, R.A. and Shapiro, Z.Ya. (1963)
             Representations of the Rotation and Lorentz
             Groups and their Applications Pergamon,
             Oxford.

36. Geroch, R.P. (1967) "Topology in general relativity." J.Math.
             Phys. 8,4, 782-786.

37. Geroch, R.P. (1968) "Spinor structures of space-times in
             general relativity I." J.Math.Phys. 9,11,
             1739-1744.

38. Geroch, R.P. (1970) "Spinor structures of space-times in
             general relativity II." J.Math.Phys. 11,1,
             343-348.

39. Geroch, R.P. (1971) "Space-time structure from a global
             viewpoint." In Sachs, R.K. Ed. General
             Relativity and Cosmology (Proc. 1969 Int.
             School Enrico Fermi) Academic Press, New
             York, pp. 77-103.

40. Greub, W., Halperin, S. and Vanstone, R. (1972) <u>Connections,</u>
                        <u>Curvature and Cohomology I</u> Academic Press,
                        New York.

41. Hájíček, P. and Schmidt, B.G. (1971) "The b-boundary of
                        tensor bundles over a space-time" Commun.
                        Math.Phys. 23, 285-295.

42. Hano, J. and Ozeki, H. (1956) "On the holonomy groups of
                        linear connections." Nagoya Math.J. 10,
                        97-100.

43. Hawking, S.W. and Ellis, G.F.R. (1973) <u>The Large Scale</u>
                        <u>Structure of Space-time</u> C.U.P., Cambridge.

44. Hennig, J.B. (1978) "G-structures and space-time geometry I,
                        geometric objects of higher order."
                        Preprint IC/78/46 International Centre for
                        Theoretical Physics, Trieste.   (Part II in
                        preparation.)

45. Hermann, R. (1970) <u>Vector Bundles in Mathematical Physics I</u>
                        Benjamin, New York.

46. Herrlich, H. and Strecker, G.E. (1973) <u>Category Theory</u>
                        Allyn and Bacon, Boston.

47. Higgins, P.J. (1971) <u>Categories and Groupoids</u> Van Nostrand
                        Reinhold, London; Mathematical Studies 32.

48. Higgins, P.J. (1974) <u>Introduction to Topological Groups</u>
                        C.U.P., Cambridge; L.M.S. Lecture Notes 15.

49. Hirsch, M.W. (1976) <u>Differential Topology</u> Springer-Verlag,
                        New York.

50. Hochschild, G. (1965) <u>The Structure of Lie Groups</u> Holden-
                        Day, San Francisco.

51. Hodge, W.V.D. (1952) <u>Theory and Application of Harmonic</u>
                        <u>Integrals,</u> C.U.P., London.

52. Husemoller, D. (1975) <u>Fibre Bundles</u> Springer-Verlag, New
                        York (2nd Edition).

53. Ihrig, E. (1976) "The holonomy group in general relativity
                        and the determination of g_{ij} from T^i_j."
                        GRG, 7, 313-323.

54. Jameson, G.J.O. (1974) <u>Topology and Normed Spaces</u> Chapman
                        and Hall, London.

55. Johnson, R.A.   (1977) "The bundle boundary in some special
                    cases." J.Math.Phys. 18, 898-902.

56. Kobayashi, S. (1972) <u>Transformation Groups of Differential
                    Geometry</u> Springer-Verlag, Berlin.

57. Kobayashi, S. and Nomizu, K. (1963) <u>Foundations of
                    Differential Geometry I</u> Interscience, New
                    York.

58. Kobayashi, S. and Nomizu, K. (1969) <u>Foundations of
                    Differential Geometry II</u> Interscience, New
                    York.

59. Kowalsky, H-J. (1965) <u>Topological Spaces</u> Academic Press,
                    New York.

60. Lang, S. (1972) <u>Differential Manifolds</u> Addison Wesley,
                    Reading, Massachussets.

61. Lee, C.W. (1976) "A restriction on the topology of Cauchy
                    surfaces in general relativity." Commun.
                    Math.Phys. 51, 157-162.

62. Lee, C.W. (1978) "Topology change in general relativity."
                    Proc.Roy.Soc. A364, 295-308.

63. Lee, K.K. (1973) "Global spinor fields in space-time."
                    GRG 4, 421-433.

64. Lee, K.K. (1975) "On the fundamental groups of space-times
                    in general relativity." GRG 6, 239-242.

65. MacDonald, I.D. (1968) <u>The Theory of Groups</u> Clarendon Press,
                    Oxford.

66. MacLane, S. (1968) <u>Geometrical Mechanics</u>, Mathematics
                    Department Lecture Notes, Univeristy of
                    Chicago.

67. MacLane, S. (1971) <u>Categories for the Working Mathematician</u>
                    Springer-Verlag, New York.

68. Maddox, I.J. (1970) <u>Elements of Functional Analysis</u> C.U.P.,
                    Cambridge.

69. Marathe, K.B. (1972) "A condition for paracompactness of a
                    manifold." J.Diff.Geom. 7, 571-573.

70. Markus, L. (1955) "Line element fields and Lorentz structures
                    on differentiable manifolds." Annals.Math.
                    62, 3, 411-417.

71. Misner, C.W.   (1963)   "The flatter regions of Newman, Unti
                              and Tambourino's generalized Schwarzschild
                              space." J.Math.Phys. 4, 924–937.

72. Misner, C.W., Thorne, K. and Wheeler, J.A.   (1969)
                              Gravitation Freeman, San Francisco.

73. Munkres, J.R.   (1954) Elementary Differential Topology
                              Princeton University Press, Princeton.

74. Palais, R.S.   (1968) Foundations of Global Non-Linear
                              Analysis Benjamin, New York.

75. Penrose, R.   (1968) "The structure of space-time." In C.M.
                              DeWitt and J.A. Wheeler, Eds. Battelle
                              Rencontres in Mathematics and Physics
                              Benjamin, New York; pp. 121–235.

76. Pirani, F.A.E.   (1965) in Lectures in General Relativity
                              Brandeis Summer Institute in Theoretical
                              Physics, 1964. Prentice Hall, Eaglewood
                              Cliffs, N.J.

77. Pontryagin, L.S.   (1966) Topological Groups Gordon and Breach,
                              New York (2nd edition).

78. Porteous, J.R.   (1969) Topological Geometry Van Nostrand
                              Reinhold, London.

79. Poston, T. and Stewart, I.   (1978) Catastrophe Theory and its
                              Applications Pitman, London (2nd edition).

80. Radivoiovici, M.   (1979) "On the geometry of the tangent
                              bundle of order 2." Contribution at Bolyai
                              Society Conference; to appear in Proceedings
                              (1980).

81. Sachs, R.K.   (1973) "Spacetime b-boundaries." Commun.Math.
                              Phys. 33, 215–220

82. Sachs, R.K. and Wu, H.   (1976) General Relativity for
                              Mathematicians Springer-Verlag, New York.

83. Schmidt, B.G.   (1968) Riemannsche Räume mit mehrfach transitiver
                              Isometriegruppe Doktorarbeit, University of
                              Hamburg.

84. Schmidt, B.G.   (1971) "A new definition of singular points in
                              general relativity." GRG 1, 269–280.

85. Schmidt, B.G.   (1973)   "Local completeness of the b-boundary."
               Commun.Math.Phys. 29, 49-54.

86. Schmidt, B.G.   (1974)   "A new definition of conformal and
               projective infinity of spacetimes."   Commun.
               Math.Phys. 36, 73-90.

87. Schwartz, J.T.   (1968)   Differential Geometry and Topology
               Gordon and Breach, New York.

88. Serre, J-P.   (1965)   Lie Algebras and Lie Groups   Benjamin,
               New York.

89. Simms, D.J. and Woodhouse, N.M.J.   (1976)   Geometric
               Quantization   Springer-Verlag, Berlin;
               Lecture Notes in Physics 53.

90. Singer, I.M. and Thorpe, J.A.   (1967)   Lecture Notes on
               Elementary Topology and Geometry,   Scott,
               Foresman, Illinois.

91. Slupinski, M.J. and Clarke, C.J.S.   (1980)   "Projective limits
               in relativity."   Commun.Math.Phys. (in press).

92. Spanier, E.   (1966)   Algebraic Topology   McGraw-Hill, New York.

93. Spivak, M.   (1965)   Calculus on Manifolds   Benjamin, New York.

94. Spivak, M.   (1975)   Differential Geometry Vol. I   Publish or
               Perish Inc. Brandeis University.

95. Steen, L.A. and Seebach Jr., J.A.   (1978)   Counterexamples in
               Topology   Springer-Verlag, New York (2nd
               edition)

96. Steenrod, N.   (1955)   Topology of Fibre Bundles   Princeton
               University Press, Princeton.

97. Strooker, J.R.   (1978)   An Introduction to Categories,
               Homological Algebra and Sheaf Cohomology.
               C.U.P., Cambridge.

98. Tipler, F.J.   (1977)   "Singularities and causality violation."
               Annals.Phys. 108, 1-36.

99. Tipler, F.J., Clarke, C.J.S. and Ellis, G.F.R.   (1978)
               "Singularities and horizons"   Preprint
               University of California to appear in the
               GRG Einstein Centenary Volume.

100. Wall, C.T.C.   (1972)   A Geometric Introduction to Topology
               Addison Wesley, Reading, Massachussetts.

101. Weyl, H.   (1922)   <u>Space Time Matter</u>   Dover, New York.

102. Whiston, G.S.   (1974)   "Hyperspace cobordism theory of space-
             time."   Int.J.Theor.Phys. 11, 285-288.

103. Whiston, G.S.   (1974)   "Topics on space-time topology II."
             Int.J.Theor.Phys. 11, 341-351.

104. Whiston, G.S.   (1974)   "Topics on space-time geometry."   GRG,
             5, 525-537.

105. Whiston, G.S.   (1975)   "Topics on space-time topology III"
             Int.J.Theor.Phys. 12, 225-240.

106. Whiston, G.S.   (1978)   "Compact spinor space-times."   J.Phys.
             A. 11, 1203-1209.

107. Yano, K. and Ishihara, S.   (1973)   <u>Tangent and Cotangent</u>
             <u>Bundles</u>   Dekker, New York.

Supplementary Bibliography

108. J.K.Beem and P.E.Ehrlich. (1981) Stability of geodesic completeness
    for Robertson-Walker spacetimes.Gen.Relativ.Grav.13,239-255.

109. J.K.Beem and P.E.Ehrlich. (1981) Global Lorentzian Geometry
    Marcel Dekker,New York.

110. J.K.Beem and P.E.Ehrlich. (1987) Geodesic completeness and stability.
    Math.Proc.Camb.Phil.Soc.102,319-328.

111. J.K.Beem and P.E.Parker. (1983) Klein-Gordon solvability and the
    geometry of geodesics.Pacific J.Math.107,1-14.

112. J.K.Beem and P.E.Parker. (1985) Whitney stability of solvability.
    Pacific J.Math. 116,11-23.

113. D.Canarutto. (1987) An introduction to the geometry of singularities
    in general relativity, Preprint Inst.Mat.App.G.Sansone, Florence.

114. D.Canarutto and C.T.J.Dodson. (1985) On the bundle of principal
    connections and the stability of b-incompleteness of manifolds.
    Math.Proc.Camb.Phil.Soc. 98, 51-59.

115. D.Canarutto and P.W.Michor. (1986) On the stability of
    b-incompleteness in the Whitney topology on the space of connections.
    Preprint,Inst.Mat.App. G.Sansone, Florence.

116. C.J.S.Clarke and A.Królak. (1985) Conditions for the occurrence of
    strong curvature singularities. J.Geom.Phys.2 N2, 127-143.

117. L.A.Cordero, C.T.J.Dodson and M.deLeon. (1988) Differential Geometry
    of Frame Bundles D.Reidel,Dordrecht.

118. M.deLeon and P.R.Rodrigues. (1985) Generalised Classical Mechanics
    and Field Theory. North Holland, Notas de Mathematica 102,
    Amsterdam.

119. L.DelRiego and C.T.J.Dodson. (1988) Sprays, universality and stability.
    Math.Proc.Camb.Phil.Soc. 103 (1988) in press.

120. C.T.J.Dodson. Systems of connections for parametric models.
    In,Proc.Workshop on Geometrization of Statistical Theory Lancaster
    28-31 October 1987, ed.C.T.J.Dodson, ULDM Publications,
    Univ.Lancaster 1987. pp 153-170.

121. C.T.J.Dodson and M.Modugno. (1984) Connections over connections and

a universal calculus. Proc.Vl Convegno Nazionale de Relativita
Generale e Fisica della Gravitazione. Florence, 10-13 October. 1984.

122. G.F.R.Ellis and B.G.Schmidt. (1977) Singular spacetimes. Gen. Relativity
      and Grav. 8,11,915-953.

123. G.F.R.Ellis and B.G.Schmidt. (1979) Classification of singular spacetimes.
      Gen. Relativity and Grav. 10,2,989-998.

124. P.L.García. (1972) Connections and 1-jet fibre bundles . Rend.Sem.Mat.
      Univ.Padova 47,227-242.

125. M.J.Gotay and J.A.Isenberg. (1980) Geometric quantization and
      gravitational collapse.Phys Rev.D 22,235-260.

126. D.E.Lerner.The space of Lorentz metrics.Commun.Math.Phys.32,19-38.

127. L.Mangiarotti and M.Modugno.(1983) Fibred spaces,jet spaces and
      connections for field theories.In Proc. International Meeting on
      Geometry and Physics ,Florence, 12-15 October 1982,
      ed.M.Modugno,Pitagora Editrice, Bologna,pp 135-165.

128. P.Michor.(1980) Manifolds of Differentiable Mappings Shiva,Orpington.

129. M.Modugno.(1986) An introduction to systems of connections.Preprint,
      Inst.Mat.App.G.Sansone,Florence.

130. M.Modugno.(1987) Linear overconnections.Preprint,
      Inst.Mat.App.G.Sansone,Florence.

131. M.S.Narasimhan and S.Ramanan. (1961) Existence of universal
      connections I.Amer.J.Math.83,563-572.

132. M.S.Narasimhan and S.Ramanan. (1963) Existence of universal
      connections II.Amer.J.Math.85,223-231.

133. D.J.Simms and N.M.J.Woodhouse. (1976) Lectures on Geometric Quantization.
      Lecture Notes in Physics 53, Springer-Verlag,Berlin.

134. F.Treves.(1980) Introduction to Pseudo-differential and Fourier Integral
      Operators Vol I:Pseudo- differential Operators. Plenum Press,New York.

135. J.A.G.Vickers. (1980) Quasi-regular singularities in general relativity.
      D.Phil. Thesis,University of York.

136. J.A.G.Vickers. (1985) Unstable singularities in general relativity.
      Class.Quantum Grav.2,755-773.

137. P.M.Williams. (1984) Completeness and its stability on manifolds
      with connection.Ph.D.Thesis, University of Lancaster.

Index